Heating Technology

Principles, Equipment, and Application

Second Edition

Heating Technology
Principles, Equipment, and Application

Second Edition

S. Don Swenson
North Seattle Community College

Delmar Publishers

I(T)P An International Thomson Publishing Company

Albany • Bonn • Boston • Cincinnati • Detroit • London • Madrid • Melbourne • Mexico City
New York • Pacific Grove • Paris • San Francisco • Singapore • Tokyo • Toronto • Washington

NOTICE TO THE READER

Delmar Staff:

Acquitions Editor: Vern Anthony
Project Editor: Patricia Konczeski

Art/Design Coordinator: Heather Brown
Production Coordinator: Karen Smith

COPYRIGHT © 1995
by Delmar Publishers
A division of International Thomson Publishing Inc.
The ITP logo is a trademark under license

Printed in the United States of America.

For more information contact
Delmar Publishers
3 Columbia Circle, Box 15015
Albany, New York 12203–5015

International Thomson Publishing Europe
Berkshire House 168-173
High Holburn
London, WC1V7AA
England

Thomson Nelson Australia
102 Dodds Street
South Melbourne 3205
Victoria, Australia

Nelson Canada
1120 Birchmount Road
Scarborough, Ontario
Canada M1K 5G4

International Thomson Editores
Campos Eliseos 385, Piso 7
Col Polanco
11560 Mexico DF Mexico

International Thomson Publishing GmbH
Konigswinterer Strasse 418
53227 Bonn
Germany

International Thomson Publishing Asia
221 Henderson Road
#05-10 Henderson Building
Singapore 0315

International Thomson Publishing-Japan
Hirakawacho-Kyowa Building, 3F
2-2-1 Hirakawacho
Chiyoda-ku, Tokyo 102
Japan

10 9 8 7 6 5 4 3 2 1 XX 99 98 97 96 95 94

Library of Congress Cataloging-in-Publication Data

Swenson, S. Don, 1928-
    Heating technology: principles, equipment, and application /S.
Don Swenson. — 2nd ed.
      p.   cm.
    Includes index.
    ISBN 0–8273–6411–3
    1. Heating.   I. Title.
TH7222.S94 1995
697—dc20
                              94–16901
                                CIP

Contents

Chapter 10 Hydronic Heating Systems and Equipment.........................115

Chapter 11 Unit Controls........................128

Chapter 12 Control Systems 145

Chapter 13 Heat Loss Variables and Factors 155

Chapter 14 Calculating Heat Loads.....172

Chapter 15 Selection and
Sizing of Heating Equipment 195

Chapter 16 Warm Air
Distribution and Duct Sizing...............212

Chapter 17 Warm Air
Registers and Grilles..............................232

Chapter 18 Air Balance 243

Chapter 19 Hydronic System Components...263

Chapter 20 Hydronic System Design and Component Sizing............276

Chapter 24 Heat Pump Sizing and Application

Preface

The first edition of "Heating Technology: Principals, Equipment, and Application" has obviously been used in many classrooms, and by many students, to gain a knowledge of the fundamental processes involved in heating a building for comfort, and also of the equipment, and application of that equipment to achieve a level of comfort in a building.

With the passage of time much new technology has been developed in regards to heating. Burners, heat exchangers and controls have all been designed to provide more efficient combustion and heating processes. These should all be considered as part of the heating processes currently used, and as such are included in this second edition. But at the same time, basic combustion processes and the design of basic heating systems must still be understood by a heating technician.

The production and management of heat is one of the most important functions of a total comfort system for a residential or commercial building. Most people take these elements for granted. They seldom realize that the level of comfort in a building depends on the knowledge and skills of trained heating technicians. Heating technicians must understand the basic processes of heat production, the operation of heating equipment, and the principles of system design. They must be able to calculate the heating requirements of a building, choose the correct equipment for the application and design of an efficient distribution system, and must be able to analyze the operation of an entire system once it is in place.

Heating Technology: Principles, Equipment, and Application provides a practical and technical sound textbook and reference book on heating for both practicing and future heating technicians. The book covers fundamental principles and skills for all common heating systems in a single volume, including coverage of the most modern equipment and systems used.

The basics-physics, fuels, combustion, and heating mediums are covered, with the important types of warm air and hydronic heating equipment included. Heat load calculations and equipment selection are covered next. Warm air and hydronic system design are covered next. Chapters on troubleshooting of a warm air system have been expanded as have the chapters on heat pump systems . Most of the information in this text is related to small heating units and systems; however, the same principles apply to larger units and systems.

Readers will find several features in this second edition that make both the first reading and later review or reference easy. New words and technical terms are defined in the margin of the page where they first appear. Each chapter concludes with a summary of important points and an instructors guide is available with numerous self-study questions. Finally, reference materials that are frequently used on the job are gathered in Appendix Tables at the end of the book.

ACKNOWLEDGMENTS

The author would like to express his thanks to Greg Skudlarek, Minneapolis Technical College, Robert F. Burns, Riverside Community College, and Walt Kirschman, Bellingham Technical College, for reviewing the manuscript of *Heating Technology: Principles, Equipment, and Application*.

The author would also like to thank the following manufacturers and photographers who provided photographs and illustrations for the preparation of the book:

Adams Manufacturing Company
Air Conditioning and Refrigeration Institute
Air Conditioning Contractors or America
Airtherm Manufacturing Company
Alco Controls Division of Emerson Electric Company
Alnor Instrument Company
American Air Filter Sales
American Gas Association
American Society of Heating, Refrigeration and
    Air-Conditioning Engineers
ARCO Comfort Products Company
Bacharach Instrument Company
Barry Blower, a Marley Company
R.W. Beckett Corporation
Bell and Gossett ITT
Bryant, Day & Night, Payne
Burnham Corporation, Hydronics Division
Dwyer Instruments, Inc.
Emerson Electric
Farr Filter Company

Friedrich Air Conditioning and Refrigeration
    Company, Room Air Conditioning Division
General Electric Company
Honeywell, Inc.
ITT Fluid Handling Division
Johnson Controls, Inc., Control Products Division
Kewanee Boiler Corporation
Lennox Industries, Inc.
Lima Register Company
National Environmental Systems Contractors
    Association
Pearless of America, Inc.
Perfection Products Company
Rheem Manufacturing Company
Robertshaw Controls Company
Taylor Instrument/Sybron Corporation
York Division of Borg-Warner Corporation
United Technologies, Bacharach

1.1 INTRODUCTION

This text is written primarily for students interested in heat as it is used to keep buildings comfortable. The physical composition of material is important to an understanding of heat and temperature. In this chapter, the molecular theory of material composition and the effect of heat is covered. The chapter covers heat and temperature and the relationship between them. Heat and temperature are defined, and the methods used for measuring them are covered. Heat transfer and the terms used in relation to heat transfer also are described and explained.

1.2 MATTER

Scientists believe that matter is made up of latticelike structures called molecules, which in turn are made up of atoms. An atom is described as the smallest particle an element can be divided into and still exist as an element. A molecule is described as the smallest particle that a combination of elements can be divided into and still retain the properties of a compound or given material. Atoms are made up of smaller particles, or bundles of energy, called **neutrons, protons,** and **electrons.** The atom is the basic building block of material.

Neutrons and protons make up the nucleus of the atom. Orbiting about this nucleus are other particles called electrons. Figure 1-1 illustrates the construction of an atom. The number of particles forming the nucleus of an atom, and the number of electrons surrounding the nucleus, determine the element formed by the atom. The types of atoms that combine in a molecule determine the material formed. If all of the atoms are of the same kind, the material is an **element.** If atoms of different kinds combine, the molecules and the material are called a **compound.** When molecules combine to form elements or compounds, they do so in a latticelike arrangement (see Figure 1-2).

In a pure element, all of the atoms are alike. More than 100 elements are now known. Most of these are natural elements, such as carbon, hydrogen, and oxygen, but some (including nobelium, law-rencium, and mendelevium) have been created by physicists in their work with nuclear

CHAPTER

1

Heat

Molecule*

Atom – The smallest particle of an element that can take part in a chemical reaction and retain its identity.

Neutron – A basic part of the atom; particles with a neutral polarity that are part of the nucleus.

Proton – A basic part of the atom; particles with a positive polarity that are part of the nucleus.

Electron – A basic part of an atom; particles of energy with a negative electric charge that rotate in orbit around the nucleus.

*Term defined in text.

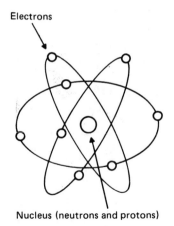

FIGURE 1–1_____
Basic parts of an atom. Neutrons and protons are found in the nucleus, and electrons orbit the nucleus.

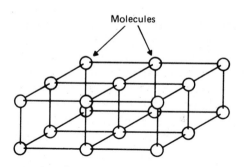

FIGURE 1–2 _____
Molecule. Molecules combine in a lattice arrangement to form an element or compound.

Element*

Compound*

Cohesion*

Heat*

physics. The term *created* is used loosely here since matter cannot be created; instead, the term means that the basic parts of an atom can be reorganized to form new elements.

In cases where atoms of two or more kinds combine to form molecules, the resulting material is a compound. An example of a compound is table salt. Table salt is made up of atoms of sodium and atoms of chlorine that have combined and then formed molecules. The molecules of the two materials combine to form sodium chloride, or common table salt (see Figure 1-3).

FIGURE 1–3 _____
Compound. Atoms of different elements combine to form compounds.

By definition, an element is a material in which all of the atoms are the same, and the material cannot be divided into particles smaller than an atom without the material losing its identity as an element. A compound is a material made up of atoms of two or more kinds; the material cannot be subdivided into particles smaller than molecules without losing its identity as a compound. In other words, salt can be divided into the elements of sodium and chlorine, but the results will no longer be salt.

1.3 DEFINITION OF HEAT

In the composition of matter, molecules combine in lattice-like arrangements. These molecules are held together by mutual attraction, friction, and pressure, which we call **cohesion.**

Matter often appears solid and rigid in form. This could lead us to think that the arrangement of molecules making up the material is static or unmoving. Actually, the molecules in any material are moving. Even iron and steel, which feel so solid, are made up of molecules that are constantly in motion. The molecules have enough cohesion to keep the materials within certain boundaries that define the shape of the material, but the molecules do move within the material.

The amount of motion of the molecules indicates the amount of heat in the material. When any material is heated, the relative motion of the molecules increases. Thus, the amount of motion of the molecules indicates the amount of heat that has been added (see Figure 1-4).

FIGURE 1–4

Molecular motion. Relative motion of molecules in a material indicates the amount of heat in the material.

EXAMPLE 1-1

When heat is added to ice, the ice changes into water. When more heat is added, the water changes to steam. Here we observe the change from solid to liquid and to gas. The motion of the molecules increases as heat is added. See Figure 1-5 for an illustration of this law of physics.

Heat can be defined as the measurement of the amount of motion of the molecules in matter. This definition is not a very practical one for application, but it does help to explain what is observed.

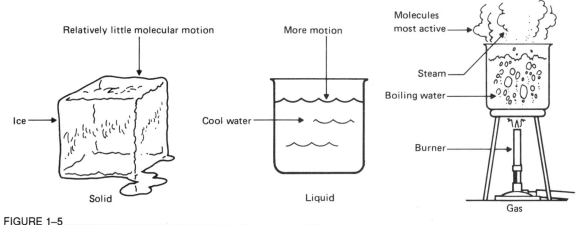

FIGURE 1–5

Temperature and state of a material. Both are related to the speed at which the molecules are moving.

Heat and State of Matter

Matter exists in one or more of three forms or states: solid, liquid, or gas.

In the solid **state,** the motion of the molecules is somewhat limited. The molecules will stay within certain boundaries without special containment. An example of a substance in the solid state is ice, the solid form of water.

In the liquid state, the motion of the molecules has increased, and they will no longer stay within definite boundaries unless they are in a container that limits their motion. In a liquid, gravity exerts enough force downward on the molecules to prevent them from moving up, with the exception of those that escape from the surface by evaporation. A container with sides and bottom is necessary to prevent the molecules from moving down or out to the sides. Figure 1-6 illustrates molecular action in a liquid. Water in a glass or pitcher is a good example of the behavior of molecules when a material is in a liquid state.

In the gaseous state, molecules are moving so rapidly that they have a tendency to dissipate in all directions. Consequently, a gas expands to fill all the space in any container in which it is placed. Steam from boiling water is a good example of a gas. The steam will fill the space over boiling water, such as the top of a teakettle, and will exert equal pressure on all surfaces of the container, including the surface of the water.

FIGURE 1–6

Molecular action in a liquid. The molecules have enough speed to allow the material to move out and down, but not enough to break surface film.

The speed at which the molecules in any material moves is increased when heat is applied. Consequently, heat added to ice changes the ice to water; more heat added to the water changes the water to steam. The change has been from solid to liquid to gas. The application of heat is necessary to increase the speed of the molecular motion. Conversely, the relative motion of the molecules indicates the heat content of the material. Figure 1-4 illustrates this principle.

At normal atmospheric pressure, water can stay in the solid state only when its temperature is 32°F or lower. If heat is added to raise the temperature above 32°F, the ice melts, or changes into the liquid state. The heat involved in this change is called **latent heat.** Heat that causes a change of state without a change in the temperature that can be measured with a thermometer is latent heat. Water between 32°F and 212°F remains in the liquid state. Heat that changes the measurable temperature of a material but does not change the state is called **sensible heat.** This term is used because the temperature difference can be sensed. It also can be measured with a thermometer. If the temperature is increased above 212°F, the water turns into steam, the gaseous state. Each increase in temperature is represented by a definite amount of heat that has been added to the water. Figure 1-7 is a graph showing the heat-temperature relationship for water in the three states.

Transfer of Heat

To use heat as a means of controlling the temperature in a space, heat has to move from the point of generation to a point of utilization. To move the heat, one or more of the methods of heat transfer is used. The methods are

1. Conduction
2. Convection
3. Radiation

Heat is moved by all three methods.

The second law of **thermodynamics** states that heat flows from a warm body to a cooler one, *never* in the opposite direction unless energy is used to move it.

When heat is transferred by **conduction,** heat flows from one molecule to another in the material being heated. An example is a bar of iron with one end held in a flame.

State – Form, or physical structure, of a material, such as solid, liquid, or gas.

Latent heat*

Sensible heat*

Thermodynamics – The science of heat and its related phenomena.

FIGURE 1–7
Graph representing water in three states and heat related to the temperature for 1 lb.

FIGURE 1–8 _____
Conduction. Heat is transferred through bottom of pan.

The end of the bar that is in the flame gets very hot; as the heat travels along the bar it also gets hot. Conducted heat is not transferred instantaneously because it takes time for each molecule to be heated in turn and then pass the heat along to the next molecule. Figure 1-8 illustrates this principle. Heat from a stove burner is conducted through the bottom of a pan to boil water in the pan.

When heat is transferred by **convection,** the heat is carried by the molecules in the material that is heated. An example is the heat carried by the air from a heating coil. The molecules of air carry the heat. Another example is heat carried by water that is heated and then pumped through a piping system to remote locations. In these examples, heat is being carried by the air or water (see Figure 1-9).

The third method of heat transfer, **radiation,** depends on heat waves or rays to transfer the heat. Heat waves function similarly to light waves. If two bodies of different temperatures are reasonably close to each other, heat will flow from the warmer to the cooler, through any intervening space, by radiation. An example is the heat we feel from an open fire. If we stand close to a fire and feel warm on the side facing the fire, our other side may still feel cold. Radiation heat can act only in direct line of sight. Any intervening body will intercept the waves and be heated by them, but the heat will not go beyond that body. Radiant heat does not warm the air directly; it warms only the objects that the waves themselves reach. Figure 1-10 shows an example of radiant heat transfer.

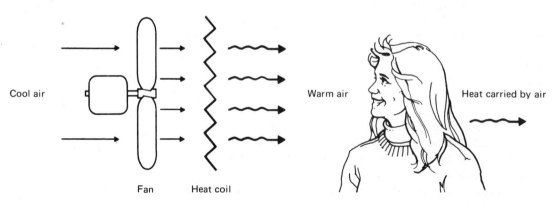

FIGURE 1–9_____
Convection. Heat is transferred as molecules of air move.

Measurement of Heat

Heat is a term related to quantity. One of the most common terms used in relation to heat is **temperature,** which is a term related to the intensity of heat and only indirectly related to the quantity of heat. To calculate heat as a quantity, we first measure temperature.

Temperature.   Temperature is a measurement of the intensity of heat. When the temperature of a room is 76°F, this reading indicates the intensity of heat in the room. If more heat is introduced into the room by the heating system, the temperature will increase, and if heat is lost from the room, the temperature will decrease. In these cases, the temperature of the air is related to the quantity of heat involved; but it is not a direct measurement of the quantity of heat.

Thermometers.   The most common device for measuring temperature is the **thermometer.** The operation of most thermometers is based upon the principle that materials expand as they are heated and contract as they cool.

The standard glass thermometer is a closed glass tube with a small-diameter bore and a glass bulb at the bottom. Figure 1-11 shows a typical glass thermometer.

The bulb and some portion of the tube contain a liquid with a low freezing point, such as alcohol or mercury. As the temperature surrounding the glass bulb increases, the alcohol or mercury expands, and the column of liquid moves up in the stem of the thermometer. If the temperature goes down, the alcohol or mercury contracts, and the column goes down because the volume of the liquid is reduced. The glass tube is marked in equally spaced gradations over a certain length called the scale. The intensity of the heat surrounding the thermometer can be read by referring to the movement of the column of liquid on the particular scale being used.

Many types of thermometers are available, including direct-reading dial types that are actuated by bimetal springs and remote-reading dial types that are actuated by thermocouples and thermistors. A thermocouple is an instrument with two wires of different types of metal twisted together. This is called a couple. A chemical reaction between the two wires generates a low-voltage electrical potential between the wires. This potential varies as the temperature at the

FIGURE 1–10 _____
Radiant heat transfer. Heat is carried by rays of energy.

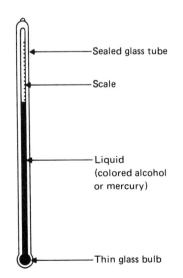

FIGURE 1–11 _____
Standard glass thermometer

Dial-Type of Thermometer

Pocket-Type Glass Thermometer

FIGURE 1–12
Two types of thermometers commonly used by heating technicians

point where the two wires are twisted together. The other ends of the two wires are connected to a magnetic coil in an indicating device. As the voltage in the wires varies with temperature, a needle on a temperature scale moves, and thus the temperature relative to the location of the couple is indicated. A thermistor is an electronic device that controls the voltage in a circuit in response to temperature at the thermistor. Figure 1-12 shows some examples of thermometers.

Special thermometers are available to measure extremely high temperatures, such as those found in flames, and to measure low temperatures, such as those found in refrigeration systems.

Temperature Scales. The **Fahrenheit** scale has been the standard scale for measuring temperature in the United States. This scale originally was established by using the temperature of an ice–salt mixture as 0°F, and the temperature of boiling water at sea level as 212°F. This scale places the temperature of freezing water at 32°F.

The other common scale used on thermometers is called the **Celsius**. The Cêlsius scale is used in the metric system. The Celsius is the common scale used for temperature measurements in all countries of the world except the United States, but it also is used in the United States for scientific and engineering work. The Celsius scale is now being adopted in the United States along with other metric values.

In the Celsius scale, the freezing point of the water at sea-level pressure is 0C, and the boiling-point temperature at the same pressure is 100C. See Figure 1-13 for an illustration of the two common temperature scales.

In solving application problems, it often is necessary to convert Fahrenheit temperature readings to Celsius, and vice versa. To do this conversion, the relationship of the Fahrenheit scale to the Celsius scale can be stated as a

FIGURE 1–13
Fahrenheit and Celsius thermometers

mathematical ratio. On the Fahrenheit scale, 180° separates the freezing point of water subtracting 32° from 212°.

$$212° - 32° = 180°$$

On the Celsius scale, 100° separates the freezing point of water (0°), as found by subtracting 0° from 100°.

The ratio of the Fahrenheit scale to the Celsius scale, then, is 180:100, or 1.8:1. This ratio gives a factor of 1.8, which is used in the conversion process. The 32° difference between the freezing point temperatures on the two scales also has to be considered.

To convert a Fahrenheit temperature reading to a Celsius, the difference between the freezing points (32°) is subtracted from the Fahrenheit reading, and the remainder is divided by 1.8 (the ratio between the two scales), as follows:

$$C° = \frac{F° - 32°}{1.8}$$

EXAMPLE 1-2

If the temperature of the air leaving an air-conditioning coil is 59°F, what is the Celsius temperature?

Solution

$$C° = \frac{59° - 32°}{1.8}\ \ 15° C$$

To convert a Celsius temperature, multiply by 1.8 and add 32° to the product:

$$F° = (C° \times 1.8) + 32°$$

EXAMPLE 1-3

If the temperature of the air leaving a furnace is measured at 73C, what is the Fahrenheit temperature?

Solution

$$F° = (C° \times 1.8) + 32°$$

$$= (73° \times 1.8) + 32° = 163.4°F$$

Technicians need to be familiar with two other temperature scales. The scales are seldom used in practical

FIGURE 1–14 _____
Comparison of the four temperature scales

application, but are used in many calculations concerning air and temperature. Both of the scales in these systems are related to **absolute zero,** the theoretical point at which no heat exists in a substance. The names of the scales are the **Rankine** and the **Kelvin,** each named after a famous scientist. The Rankine scale is related to the Fahrenheit scale and the Kelvin to the Celsius. Figure 1-14 compares the four temperature scales.

Absolute zero can be computed. It is found to be 460° below 0°F and 273° below 0°C. Since Celsius and Fahrenheit scales have different bases, absolute zero is numerically different for each of the scales, but the numbers relate to the same actual temperature. No thermometer can read as low as absolute zero, but the temperature can be calculated.

Since there is a 460° difference between the zero points on the Fahrenheit scale and the Rankine scale, to change a Fahrenheit reading to a Rankine, we simply add 460 to the Fahrenheit reading as follows:

$$R° = F° + 460°$$

EXAMPLE 1-4

If a Fahrenheit temperature reading is 73° F, what is the absolute reading on the Rankine scale?

Solution

Add 460° to 73°. The sum is the Rankine reading:

$$R° = F° + 460°$$
$$= 73° + 460° = 533°R$$

Since there is 273° difference between the zero points on the Celsius and the Kelvin scales, add 273° to a Celsius reading to change it to a Kelvin reading as follows:

$$K° = C° + 273°$$

EXAMPLE 1-5

If a Celsius temperature reading is 22.8C, what is the Kelvin temperature reading?

Solution

Add 273° to the 22.8°C. The sum of 295.8°K is the answer:

$$K° = C° + 273°$$

$$= 22.8° + 273° = 295.8°K$$

Although many types of thermometers are available, each type is calibrated in either the Fahrenheit or the Celsius scale. Each may serve one particular purpose better than another, but each does the same thing: they measure the intensity of heat.

Heat Quantity

Temperature is only one of the variables that we are concerned with in a discussion about heat. Temperature is a measure of the intensity of heat; more often, we are concerned with the quantity of heat.

To measure quantity of heat, consideration must be given to the temperature plus the mass or volume of material involved in a temperature change. For instance, a teakettle of boiling water at sea level pressure that holds 1 gallon (gal) has a temperature of 212°F. Because the volume of water and steam is not large, a large quantity of heat is not involved. A large steam boiler that drives a turbine to generate electricity, however, may have hundreds of gallons of water. If the temperature of the water and steam in the boiler is 212°F, there is a great amount of heat represented because of the large volume of water and steam. More heat is needed to raise the temperature of the water in the large boiler to the boiling point than to raise the temperature of the water in the small teakettle to the boiling point.

The term used to describe quantity of heat is **British thermal unit** (Btu). The definition of a Btu is the amount of heat required to raise the temperature of 1 pound (lb) of water 1°F, or to be more precise, from 59°F to 60°F. A Btu is approximately the amount of heat released when a common wooden match is burned. See Figure 1-15.

Specific heat is the amount of heat that will raise or lower the temperature of one pound of a material 1°F. Specific heat shown on a specific heat table is relative to 1°F and is compared to water as a standard. The specific heat of

British thermal unit*

Specific heat – A ratio of the amount of heat required to raise the temperature of 1 lb of a substance 1°F compared to the amount required for water.

FIGURE 1–15 _____
British thermal units. When a wooden kitchen match is burned, approximately 1 Btu of heat is produced.

TABLE 1–1 _____
Specific Heat of Some Common Materials

Material	Specific heat (Btu/lb/°F)
Aluminum	0.22
Carbon	0.17
Concrete, stone	0.19
Copper	0.09
Glass	0.20
Ice, 32°F	0.49
Silver	0.06
Water	1.00
Air, dry, 50°F	0.24
Steam, 220°F	0.47
Beef carcass	
Above freezing	0.68
Below freezing	0.48

Calorie*

water is 1.0 since 1 Btu is needed to raise the temperature of 1 lb of water 1°F. Each material has its own specific heat. The specific heat of air is 0.24 since 0.24 Btu is needed to raise the temperature of 1 lb of air 1°F. See Table 1-1 for a comparison of the specific heat of some common materials.

In the metric system, the term **calorie** is used for heat. A calorie is described as the amount of heat required to raise the temperature of 1 gram (g) of water 1°C. In this text, the Fahrenheit scale and the term Btu will be used.

Both mass and temperature are needed to measure heat. If the weight of any material is known, and that material undergoes a temperature change, the amount of heat involved in the process can be calculated mathematically.

The formula used for calculating the amount of heat in a process is

$$Btu = ms (T_1 - T_2)$$
where:
Btu = the total heat in the process
m = the weight of the material
s = the specific heat of the material
$(T_1 - T_2)$ = the temperature difference of the process

To find the amount of heat involved in a process, multiply the weight of the material times the specific heat of the material times the temperature difference. The result will be the amount of heat in Btu.

EXAMPLE 1-6

If 125 lb of water is at a temperature of 60°F, how much heat will be required in Btu to raise the temperature to 140°F?

Solution

Multiply 125 lb times the specific heat of water (1.0) times the temperature difference (140° - 60°):

$$Btu = ms (T_2 - T_2)$$

$$= 125 \text{ lb} \times 1.0 \times (140° - 60°) = 10,000 \text{ Btu}$$

To find the amount of heat involved if a quantity of air changes temperature, multiply the weight of the air times the specific heat of air times the temperature difference.

EXAMPLE 1-7

If 200 lb of air is brought into a building through a ventilation system when the temperature outside is 36°F, and it is necessary to heat the air to 75° (the inside temperature in the building), how much heat is required?

Solution

$$\text{Btu} = ms(T_1 - T_2)$$

$$= 200 \text{ lb} \times 0.24 \text{ Btu/lb} \times (75° - 36°)$$

$$= 1872 \text{ Btu}$$

The amount of heat required to raise the temperature of the air from 36°F to 75°F is 1872 Btu.

1.4 SUMMARY

Heat is the most important consideration in providing comfort in a building, consequently, it is a necessary subject for heating, ventilating, and air-conditioning technicians to study.

To understand how heat is produced, controlled, and applied in heating a building, it is necessary to understand how materials are formed, and how those materials are affected by heat. The molecular theory makes possible an understanding of many of the phenomena observed when dealing with heat. Heat, temperature, and change of state can all be explained by the molecular theory of matter and heat.

All materials are made up of molecules, combinations of atoms, that are in motion. The relative speed with which the molecules within a material are moving depends upon the amount of heat in the material. The amount of heat determines the temperature of the material, and the speed of the molecules also determines whether the material is in the solid, liquid, or gaseous state.

Temperature is a term related to intensity or degree, measured in degrees Fahrenheit or Celsius. Heat is a term related to quantity, measured in British thermal units (Btu). The temperature in the theoretical condition in which no heat exists is called absolute zero. Scales have been developed relative to the freezing and boiling points of water. These are the Fahrenheit and Celsius scale. On the Fahrenheit scale water freezes at 32°F and boils at 212°F. On the Celsius scale water freezes at 0°C, and boils at 100° C.

1.5 TERMS TO KNOW

molecule
atom
neutron
proton
electron
element
compound
cohesion
heat
state
latent heat
sensible heat
thermodynamics
conduction
convection
radiation
temperature
thermometer
temperature scales
Fahrenheit
Celsius
absolute zero
Rankine
Kelvin
British thermal unit
specific heat
calorie

2.1 INTRODUCTION

Heat, to maintain comfort in a building, generally is produced by one of two traditional methods, or by one of several alternative methods. The first of the traditional methods is combustion of fuel under controlled conditions. The second is the use of electricity through a resistance heater. These two methods are covered in this chapter.

The realization that the world is faced with a limited source of energy has created a demand for new technology and better use of old technology in the production of heat. Some of the more common alternative methods being used in response to this demand are covered briefly in this chapter. These alternative methods include solar energy, geothermal heat, and heat pumps.

2.2 COMMONLY USED FUELS

A defination of a fuel, for the purposes of this text, can be any material that is burned to produce heat.

The **fuels** most commonly used for the production of heat are natural gas, fuel oil, liquified petroleum gases (LPG), manufactured gases, and coal. Gas, oil, and coal are similar in chemical makeup, but their physical differences require that they be handled differently as fuels. Heating technicians must understand these differences to be able to understand the procurement, distribution, and use of the fuels.

Gas Fuels

The two most commonly used gases for fuel are natural gas and liquid petroleum gas (LPG). Some of the other gases that are used in special cases are blast furnace gas, casing head gas, oil gas, refinery gas, and sewage gas. Table 2-1 lists properties of some of the fuel gases.

Because the use of many of these gases is so limited, our discussion will deal only with natural and LPGs. An understanding of those two gases will help technicians to understand the basic use of the other gases also.

CHAPTER

2

Sources of Heat

Fuel – Any Material that is burned to produce heat.

TABLE 2–1
Properties of Some Fuel Gases

Fuel	Source	Chemical composition	Btu value
Butane	By-product of refining process; also found at wellhead	C_4H_{10}	3200–3260 Btu per cu ft
Natural gas	Gas wells and found with oil	Varies; mostly CH_4 C_2H_6 and C_3H_8	950–1150 Btu per cu ft
Propane	By-product of refining process	C_3H_8	2500 Btu per cu ft
Refinery gas	By-product of refining process	Varies; mostly propane and butane	1200–2000 Btu per cu ft
Sewage gas	Sewage disposal plants	Varies	600–700 Btu per cu ft

Combustion – The process of burning; the release of heat energy through the process of oxidation.

FIGURE 2–1
Gas burner used for lighting purposes

Natural Gas. Although natural gas is found quite commonly, the origin of it is not actually known. Because natural gas usually is found in relation to petroleum products, scientists think that oil and natural gas have a common source. Both oil and natural gas are considered to be the fossil remains of plant and animal materials.

Natural gas has been used for lighting and heating purposes for thousands of years, but only within the last century has its use been common. The ancient Chinese used natural gas from shallow wells to heat seawater for the reclaiming of salt. The original development of a natural gas well for commercial production in the United States occurred in the early 1800s in New York State. Inhabitants of the area of Fredonia, New York, were intrigued by a spring from which gas flowed that could be ignited. A young gunsmith saw the possibilities of the gas, dug a shallow well to recover it, and collected and used it for lighting and heating. From this beginning, the first commercial gas company in the United States was organized.

The first common use of natural gas was for lighting purposes. Gas was burned in a simple "fishtail" burner, and the natural incandescence from the flame provided light. Figure 2-1 depicts a typical gas burner used for lighting purposes. The fishtail burner is very simple. Because air for **combustion** does not mix with the gas except in the flame, carbon particles in the fuel burn slowly and glow while they burn. A yellow flame results that produces more light than heat.

Until the development of the electric light in the late 1800s, gas lighting was the principal use for natural gas. As electric light was perfected, the gas industry looked for other uses for its product; hence the development of gas-burning heating units.

The value of natural gas as a heating fuel was recognized early, but its widespread use had to await the development of distribution piping systems. As the distribution systems spread out from the wellheads where the gas was abundant, furnace manufacturers developed gas-burning equipment that would utilize the fuel. Natural gas quickly became the most popular of the heating fuels. The main reason is that gas is convenient; it does not have to be delivered by truck or stored on the premises. This factor, and its relative abundance, make natural gas a good fuel for commercial and industrial use as well as for domestic heating.

Natural gas to be used for commercial purposes is either produced at wells drilled for that purpose, or recovered at the wellhead of oil wells. In the past, the presence of natural gas with oil was considered a nuisance, and the gas was burned off or wasted at the wellhead. Now natural gas has become valuable in its own right, and it is recovered and used.

Major gas fields are found in many different parts of the world, including the United States. The gas from these fields is distributed by large pipelines owned by major distribution companies. Through the use of interconnected systems and local distribution systems, the gas is available in almost all areas of the United States where the population is great enough to create a market. The development of the distribution system has been the most important feature in making natural gas the most commonly used fuel. Its ready availability and convenience make it popular.

Natural gas generally is distributed by a local utility company. This company is responsible for the installation and maintenance of the pipelines, as well as ensuring a constant supply of the gas. Mains, or large gas lines, bring the gas to the property line of the consumer. The utility brings a service line from the main to the gas meter adjacent to the building where the gas will be used. The installation and maintenance of the piping within the building is the responsibility of the building owner.

Pressure regulator – A device to regulate the amount of pressure in a system.

Water gauge (wg) – A manometer or device for measuring pressure in a duct or piping system; or the pressure measured by a manometer in inches of water.

Manometer – A device for measuring low pressure in a liquid or air.

Resistance*

*Term defined in text.

Pressure in the gas main is relatively high to help distribute the gas. A **pressure regulator** at the meter reduces that pressure for distribution in the building. Regulators are also used on each appliance to reduce the pressure at the burner. The pressure in the lines is measured in inches, **water gauge** (in. wg). This measurement refers to the height, in inches, of a column of water that the pressure would support. In. wg is measured by an instrument called a **manometer.**

When a gas distribution system is designed for a particular building, the amount of gas needed for each heating unit has to be considered. The gas is measured in cubic feet per hour (cu ft/hr) and is determined by the heating capacity of each unit. Pipe is sized to deliver the cu ft/hr, but the flow rate also depends on the amount of **resistance** to the flow of the gas in the pipe. A certain amount of friction always exists between the gas and the pipe walls and fittings. The accumulated value of this friction is the resistance. To ensure the flow of gas through the pipe, the gas is kept under pressure. This is to overcome the resistance to flow.

Natural gas is colorless and odorless. The chemical composition varies with its source, but methane (CH_4O) is always a major constituent. Since methane is made up of hydrogen and carbon, the fuel is called a hydrocarbon, as are all of the other common fuels. Most natural gas contains some ethane (C_2H_6) and a small amount of nitrogen. The heating value of natural gas averages about 1000 Btus per cu ft of gas. In some cases, the value may be higher, and in some it may be lower.

Liquefied Petroleum Gas. LPGs are propane and butane, or a mixture of the two. Propane and butane are by-product gases of the petroleum industry. Because they are in a liquid state, they usually are used in areas where pipelines are not run or as a reserve source of fuel supply.

LPGs have been used as fuels since the early days of petroleum production. Both propane and butane are found in the gaseous state, but they will liquefy under moderate pressure. They are liquified and stored and transported in tanks under pressure. The LPGs can then be made available to areas where gas lines do not run. Both propane, and butane are used widely as fuels. Butane also is valuable as a source of chemicals for many other processes.

LPGs are transported from the source of supply by tank trucks or train cars especially built for that purpose. At the local distributor level, the LPGs are transferred to smaller tank trucks that carry the liquid fuel to customers. Each time the fuels are transferred from one tank to another, they are pumped in the liquid state so the pressure is maintained. The consumer also must keep the fuel in a tank under pressure, however the fuel is allowed to **gasify** as it goes through a pressure regulator when it leaves its storage container. In the appliance, LPG burns in the same way as natural gas.

The use of butane is somewhat limited by the fact that it will not gasify at temperatures lower that 32°F. Because of this it should not be used as fuel when it must be stored in a cylinder located outdoors when the temperature may drop below 32°F.

Both propane and butane are made up of hydrogen and carbon, as are other fuels. Butane has a chemical formula of $C_4H_{10}O$ with traces of other gases. Propane has a formula of C_3H_8. The heat content of the LPGs is considerably higher than that of natural gas. Butane produces 3200 to 3260 Btu per cu ft of gas burned. Propane produces approximately 2500 Btu per cu ft. Table 2-1 provides comparative values.

Fuel Oil

Fuel oil is a liquid fuel derived from petroleum. Petroleum, or crude oil, comes from the well as a complex mixture of chemicals, mostly hydrocarbons. Refineries separate the mixture into various fuels, such as gasoline, kerosene, and light and heavy fuel oils. The fuel oils themselves are separated into different grades, which are numbered. The most commonly used fuel oil for small heating units is No. 2. The heavier oils, called residual oils, are heavy, tarlike oils and are normally used in stationary power plants and ships.

Oil has been used as a heating and lighting fuel for thousands of years. The original source was from seeps, places where the petroleum naturally came to the surface of the earth. Here the oil formed in pools and could be dipped up in any container. The original oil burners were open containers of oil. The oil burned either on the surface or on a wick that was placed in the oil.

Pressure drop

Gasify – To change a liquid to a gas.

FIGURE 2–2 _____
Power oil burner commonly used in residential
or small commercial furnaces

Combustion air*

Distillation – The process of boiling, or
changing a liquid to a gas, and then condensing it back to a liquid.

The first commercial oil burners were pot burners,
open burners in which the oil burned on the surface and
combustion air mixed with the flame above the point of
combustion. Combustion air is air that provides oxygen
for the combustion process. This type of burner is effective
and is still used in very small heating units, but regulating
the amount of fuel burned is not easy, and the burner is
not very efficient.

In the mid-1800s, power oil burners were developed
that mixed combustion air with the oil and introduced the
oil-air mixture into combustion chamber under pressure.
This innovation made it possible to control the burning
rate and also produced a much more efficient flame for the
combustion of oil. Now, most oil-burning heating units
use some type of power burner, and manufacturers are
continually making improvements to increase the combustion efficiency of burners. Figure 2-2 shows a modern
power oil burner commonly used in residential or small
commercial heating units.

Crude oil, the source of fuel oil, is found in pools and
layers that lie trapped in the earth's crust. Wells are drilled
in the area of the pools, and pumps are used to raise the oil
to the surface. The crude oil is then transported to refineries.
Most of the oil is transported in pipes, but some is carried by
barge, tank car, or tank truck.

All liquid fuels are products, or by-products, of the
refining process. The way the petroleum is refined has
more to do with the fuel quality than does the source of
the crude oil. Refining consists of separating and recom-
bining the hydrocarbons of the crude oil into specialized
products like gasoline, kerosene, and fuel oil. The basic
process of the refinery is **distillation**. The hydrocarbons
are separated into different fractions or groups, having
approximately the same range of boiling points. The
lighter fractions, called distillates, are represented by gasoline, kerosene, and the lighter fuel oils. The remainder, or
residual products, are heavy fuel oils, asphalt, and tars.
The light fuel oils are used for domestic and light commercial heating. The heavier oils are used for commercial and
industrial heating.

In the modern refining process, distillation is only the
beginning. Heat and pressure are used to change the
hydrocarbon structure of the petroleum so that the crude

FIGURE 2–3 —————————————————————————————————————
Some of the products of the refining process. Crude oil is separated into lighter and more volatile products.

oil will yield more of the lighter products and proportionately less of the heavier ones. A diagram showing some of the products of refining is found in Figure 2-3.

Fuel oil is stored in large tanks at the refinery until it is taken by tank trucks or tank cars to the distributing companies. For local use, dealers deliver the fuel oil in trucks to customers. The oil is stored in tanks on the site where it will be used. These tanks may be located either above or below ground. The oil is pumped from the tanks to the heating units to be fired. Lighter oils, No. 1 and No. 2, can be pumped without any processing as long as they are stored where the temperature does not fall below 0°F. When the heavier fuel oils are used, heaters have to be installed in the tank and around the oil lines to ensure an adequate flow of oil to the burner.

Fuel oil is made up mainly of carbon and hydrogen, with approximately 85 percent carbon and 15 percent hydrogen. Fuel oil also contains traces of oxygen, nitrogen, and sulfur. No. 2 fuel oil, the most common grade for residential or light commercial heating, produces approximately 140,000 Btu of heat per gal burned. When No. 2 fuel oil is used in a properly adjusted burner, the combustion efficiency is 80 percent, or 112,000 Btu per gallon; that is, the usable heat value will be 112,000 Btus per gallon of oil burned. Table 2-2 gives some characteristics and the heating value for the various grades of fuel oil.

TABLE 2–2 _____
Typical Gravity and Heating Values of Standard Grades of Fuel Oil

Grade no.	Gravity, API	Weight, lb per gallon	Heating value, Btu per gallon
1	38–45	6.95–6.675	137,000–132,900
2	30–38	7.296–6.960	141,800–137,000
4	20–28	7.787–7.396	148,100–143,100
5L	17–22	7.94–7.686	150,000–146,800
5H	14–18	8.08–7.89	152,000–149,400
6	8–15	8.448–8.053	155,900–151,300

Source: Reprinted with permission from American Society of Heating, Refrigerating, and Air-Conditioning Engineers, Inc., *ASHRAE Handbook, 1977 Fundamentals* (New York, 1977)

Coal

It is not known when coal was first used as a fuel, but remains found indicate that coal has been used for at least 4000 years. The Chinese were using coal some 3000 years ago, and the use of coal as a fuel is mentioned in the Old Testament. However, only in the last 200 years has it been used extensively.

The English, the first to see the commercial value of coal, used it widely during the 1700s to power the Industrial Revolution. Although coal was recognized in the United States by some of the earliest explorers and settlers, not until the mid-1700s was it developed for commercial use. Mining began in western Pennsylvania in 1759. Important beds of coal were discovered in eastern Pennsylvania in 1791. As the United States industrialized, the use of coal as a fuel became important. By the mid-1800s, the growth of the railroads created a direct market for coal and also enabled the distribution of coal to other places where industry could use it.

From the late 1800s to about 1950, coal was used extensively in industry and as a fuel for heating residences and commercial buildings. With the development of distribution systems for natural gas and electricity, however, the use of coal for residential heating has been almost completely abandoned. Coal continues to be important as a source of heat for commercial and industrial use. Currently, as other sources of energy become scarcer and more expensive, the use of coal is increasing in the United States.

Coal is a fairly abundant mineral in the United States and is mined commercially in many places. It is mined both underground and in open-pit mines. The mining process in the underground mines is relatively expensive and often dangerous. Consequently, the open-pit concept is used whenever the coal bed is near enough to the surface of the earth to make mining feasible. Large quantities of coal can be obtained at lower cost by the open-pit method.

Coal often is shipped directly from the mines to the point of use by direct rail lines or even by conveyor. Such is the case in many of the large generating plants, which have been built in recent years. After the initial cost of the distribution system is absorbed, this type of transportation is less expensive.

One of the greatest problems in the use of coal for fuel is that it burns "dirty." Because coal is a solid fuel, burning it generally leaves undesirable by-products. Although coal is a hydrocarbon and thus similar to natural gas and fuel oil, coal has a much higher percentage of carbon and contains other minerals. When coal is burned, in the solid form, high efficiency is difficult to achieve. Soot, clinkers (hard particles of ash), and other undesirable by-products are produced in the combustion process. These by-products must be removed, a process that increases the cost of using the fuel. In the past, when coal was used commonly as a fuel, the contamination from the smoke and soot was considered a necessary evil. But with the increased recognition of the necessity of maintaining a clean environment, preventing pollution is now considered important. Coal-burning heating equipment can be installed so that the coal burns reasonably clean, but the special equipment and processing of the fuel before burning add to the cost of the process.

The Btu content of coal varies greatly depending upon the exact chemical content of the coal. Carbon makes up the most important constituent of coal, and hydrogen and other chemicals also are found in it. Coal is broken into different grades depending on the chemical composition and the hardness. The "soft" bituminous coals found in the western part of the United States have different burning

TABLE 2–3

Chemical Composition and Approximate Heating Value of Standard Grades of Coal

Rank	Btu per Lb		Constituents, Percent					
	Moist, Mineral-matter-free[a]	Moist, as Received	Oxygen	Hydrogen	Carbon	Nitrogen	Sulfur	Ash
Anthracite......................	14,600	12,910	5.0	2.9	80.0	0.9	0.7	10.5
Semi-Anthracite...............	15,200	13,770	5.0	3.9	80.4	1.1	1.1	8.5
Low-Volatile Bituminous.......	15,350	14,340	5.0	4.7	81.7	1.4	1.2	6.0
Medium-Volatile Bituminous....	15,200	13,840	5.0	5.0	79.0	1.4	1.5	8.1
High-Volatile Bituminous A....	14,500	13,090	9.2	5.3	73.2	1.5	2.0	8.8
High-Volatile Bituminous B....	13,500	12,130	13.8	5.5	68.0	1.4	2.1	9.2
High-Volatile Bituminous C.....	12,000	10,750	21.0	5.8	60.6	1.1	2.1	9.4
Sub Bituminous B..............	10,250	9,150	29.5	6.2	52.5	1.0	1.0	9.8
Sub Bituminous C.............	9,000	8,940	35.8	6.5	46.7	0.8	0.6	9.6
Lignite.......................	7,500	6,900	44.0	6.9	40.1	0.7	1.0	7.3

[a] (Btu as received) $\times$ 100 $\div$ (100 − 1.1 Ash)

Source: Reprinted with permission from American Society of Heating, Refrigerating and Air-Conditioning Engineers, Inc., *ASHRAE Handbook, 1977 Fundamentals* (New York, 1977)

qualities and Btu content per pound than do the "hard" anthracite coals of the East. The burning properties and the amount of pollution created by the combustion of coal are determined by the hardness and the chemical composition. See Table 2-3 for the chemical makeup and heat value for the various standard grades of coal.

2.3 OTHER SOURCES OF HEAT

Fuel-burning equipment traditionally has been the most common source of heat. Other sources, however, are practical and often used. Electric heat has become as common as gas and oil heat in most parts of the United States. Solar heat and geothermal heat are both being used more commonly.

Electricity is a form of energy rather than a fuel, but when electricity is used for heating, it often is included in descriptions as a fuel. Since about 1950, the use of electricity as a source of heat has increased rapidly. Today electricity is used in new buildings as often as the older forms of fuel. In areas served by large electrical utilities, calculating the heat loss of buildings in **kilowatt per hour** (kWh), a term related to electric energy, rather than in Btu/h, a term related to heat energy, is now common practice.

Heat pumps, typically used as combination heating and cooling units, often are used for heating both residential and commercial buildings. In most cases the heat pumps utilize electricity as a source of operating energy.

Currently there is a great deal of interest in (and much development work being done with) solar energy, geothermal heat, reclaimed heat from buildings, and other sources of heat. These sources are being used experimentally, and in some cases practically, for heating. It can be expected that new **technology** related to these sources will be developed, and others found, that may change many of our traditional ways of heating.

Electrical Energy

Electricity is commonly used as a source of energy for heating. With power lines being carried into virtually all populated areas since the early 1900s, the use of electricity for heating has made a great deal of sense. In many areas where electrical power was reasonably inexpensive because of generation by water power, electricity could be used for heating purposes at less cost than more traditional fuels.

The earliest electrical experimenters recognized that heat could be produced by connecting a conducting wire to an electric battery. Whenever electric current passes through a conductor that has some resistance to the flow of the electricity, heat is produced by that resistance. When the electric light was being developed in the 1880s, the same principle was used to produce light.

Practical electric heating units have been used since the early 1900s. The first were small units that simply heated by radiation off the hot wires. During the 1920s and 1930s, systems were developed in which the wires were embedded in ceilings, floors, or walls of buildings. The surfaces became radiant panels. These systems were not particularly successful because most buildings at that time were not insulated the way they should have been for that kind of heating system.

During the 1950s, electric furnaces similar to gas or oil furnaces were developed, and electric heat could then be

Kilowatt per hour (kWh) – An electrical term related to the amount of work or heat produced by one kilowatt in one hour's time.

Technology – The science of the industrial arts.

used in circulated air systems. The development of electrical furnaces made it possible to provide all of the elements of a comfortable heating system while using the electricity as a source of heat. About the same time, electric baseboard heating and electric wall panel heaters also were developed, and their use became fairly common.

With the use of more and better insulation in buildings, electric radiation systems, with the heat in the ceilings, have become quite common. In this system, wires are embedded in the ceiling finish material, and the entire ceiling surface is heated by the wires, providing a source of radiant heat for the room below. Adequate thermal insulation must be provided above the ceiling. Although this type of heating does not provide some of the requirements of a "total comfort" system, such as circulation of air or filtration, it does have the advantage of being relatively inexpensive to install. Of the buildings that are being heated with electricity today, approximately 60 percent are heated by radiant systems and 40 percent by circulated air systems.

Electricity can be generated by several different methods. The only practical method developed so far for producing commercial quantities of electrical power is by using large dynamos for generation. This method is called **electromagnetic generation.** Electromagnetic power generation is the conversion of one kind of energy (mechanical) to another (electrical). The mechanical power to turn the dynamos for the generation of electricity is derived from waterpower, steam power, or any other reasonable and economical power source. Waterpower usually is used for the generation of electricity where such power is available, generally in the western United States.

Dams and reservoirs are found along most large rivers in the western United States. Although the water is used for irrigation, the main reason for the construction of the dams is to provide electric power generation. Water-generated power is relatively inexpensive. Where waterpower is not available, dynamos often are driven by steam turbines. The steam is produced by burning gas, coal, or oil to heat the boilers. In some areas, steam from natural geothermal sources is used for generating electricity.

The electric utility companies use interconnected distribution systems to carry the electricity from the point of generation to almost any other point in the United States. Areas so remote they do not have electric power lines are rare. Because of this extensive distribution system, electricity is a convenient energy for heating purposes.

The utility company's distribution system brings the power to the point of use; the owner provides the internal distribution system. Service lines bring the electricity from the power lines to the building. See Figure 2-4. The service lines then go through the meter at the building and to the main service panel where the disconnect and fuses, or circuit breakers, are located. A service panel with circuit breakers and branch circuit connections is shown in Figure 2-5.

At the main service panel, the power is distributed to subpanels, which serve the various branches in the building where the individual loads, such as heating units, are located. The heating system is normally on a separate branch out of a subpanel that is fused and provided with a disconnect for the heating circuit only. Since the electrical power is distributed in the main transmission lines at a high voltage, the power always is transformed down to a lower voltage for distribution to local areas.

The method of transforming determines the exact voltage and phase that will be provided to a particular building. The most common voltages are:

1. 460 or 480 volts (V), 3 phase
2. 208, 230, or 240 V, 3 phase
3. 208/12 volt, or 230/115 volt 1 phase

FIGURE 2–4
Service lines bringing in electricity from power lines into a building

Two 120-volt service lines ⟶   ← Neutral

⟵ Panel box without cover

Main breakers ⟶

230-volt circuit

120-volt circuits

120-volt branch circuits

Ground

FIGURE 2–5
Service panel used to provide branch circuits, disconnects, and overcurrent protection

When electricity is used for heating purposes, the energy conversion is 100 percent efficient since there is no loss for flue gases or combustion efficiency. The rated output of the heating equipment is 100 percent of the electrical energy input. Each kilowatt of electricity produces 3416 Btu/h of heat.

Alternative Ways to Heat

Because our supply of traditional fuels is limited and

because we are rapidly depleting them, a greatly accelerated search for new sources of fuel and energy is now going on. Scientists are looking for ways to tap some of the sources of energy that are obvious in nature but not always easy to use. Among these sources are solar energy, wind power, wave power, and geothermal heat. Not all of these alternatives have been developed to the point where their use is practical, but some are being used to some extent. All of them, plus others, will be developed over time.

Solar Heat.   Humans probably have been interested in solar heat since they first started reasoning. It is obvious that there is a great deal of heat energy in the rays that come to the earth from the sun. Although we have used this energy in some ways since before recorded history, only in the last 100 years has research and technology made it possible for us to use much of that available.

There are basically two ways to heat directly with solar energy. The first is called passive. In **a passive system,** a building itself, or parts of a building, are warmed by the sun's rays. The interior of the building is then warmed by energy stored in those parts. The other method, an **active system,** has collector cells that collect the energy from the sun's rays in a medium such as water or air, and the medium is then used to heat building spaces.

The principles of a passive solar system have been used since people started facing their buildings toward the south to catch the sunlight in the winter. Active systems are a newer technology.

In some applications solar energy is collected by photosensitive light panels and converted directly to electricity by the panels. The electricity is then used for heating and other purposes.

A complete description of the different methods of heating with solar energy will not be attempted in this text, but certainly developments of solar systems will continue and the utilization of solar energy for heating purposes will become more common.

Geothermal Heat.   The term geothermal refers to the heat found within the Earth. Scientists accept the fact that the center of the Earth is molten rock. Evidence to support this belief is seen in various places where the heat affects the

Heat pump*

Coefficient of performance (COP)*

Heating seasonal performance factor
(HSPF)

surface in such manifestations as boiling springs, geysers, and, more dramatically, in volcanoes. Under some circumstances, this heat can be tapped and utilized as a source of energy. Either the hot water that comes to the surface naturally can be used, or wells can be drilled into the earth and pumps used to bring the hot water to the surface. In many places in the world, natural hot water and steam are used as a source of heat energy for both domestic and commercial heating.

Heat pumps.   A **heat pump** is a mechanical device that moves heat from one place to another. The pump does not actually produce heat but will move the heat from a place where it is not needed to one where it is. Heat pumps make it possible for us to pick up heat from the ground, a water source, or the air and move it into a building that we want to heat. Much less energy is needed to move heat from one place to another than to produce heat by combustion or by direct use of electricity. For this reason, the heat pump is an efficient and economical heat source.

Most heat pumps utilize a refrigeration system for operation. Heat is picked up from a heat source, or heat sink, by a refrigerant. The pressure and temperature of the refrigerant are raised by compression. Then, the heat in the refrigerant is given up by an evaporation process in a coil, at a place where it can be used for heating a building. The actual commercial development of heat pumps as heating units goes back to the early 1900s, but common use dates back only to about 1950.

Most heat pumps are driven by electrical motors. But because of the principle upon which the pumps operate, the amount of heat that they put out is much more than that produced by the same amount of electricity converted directly into heat through resistance heaters. A **coefficient of performance** (COP) of 3.5 to 1 is common. The COP is the ratio of the amount of energy used to the amount of energy made available as heat. To define a term that is more indicative of annual energy use, a term, **heating seasonal performance factor** (HSPF) is used. This term is related to the seasonal performance of the heat pump in producing heat.

Wood Heating.   As a result of the worldwide energy and fuel shortages and consequent high prices, many people are using wood as a fuel to heat their homes. This use is not really new, since wood is probably the oldest fuel used, but it is new to most people in this generation.

Wood may be less expensive to purchase than other fuels, but in most cases, fireplaces or stoves used for burning it are so inefficient that much energy is wasted. Wood usually is burned in a heating unit that heats one area of a building. It can be burned in a central furnace if the furnace is designed for that purpose. Wood should be burned in an air-tight unit designed to give maximum efficiency,and then only in compliance with local or national codes relating to its use.

The actual heat value of wood as a fuel varies greatly depending upon the type of wood used, its density, and how dry it is, but the efficiency of the heating unit really determines heat output. Table A-1 in the Appendix gives the weight and heat value for some typical woods used for fuel in the United States.

2.4 SUMMARY

Traditionally the most common way of producing heat for comfort has been to use a fuel that can be burned. In the burning or combustion process, heat energy is released from the fuel. The most common of fuels, listed in the order of their development and use, is as follows: wood, coal, fuel oil, and natural gas. In recent years, electricity, which is not a fuel but a form of energy, has become important as a source of heat.

The use of the various fuels has progressed from the simple to the more complex as society's needs, and the technology to use the fuels, has changed. Wood, the first fuel used, could be burned in an open fire, but the fuels used today require complex and sophisticated equipment. The efficiency of the equipment is much better than in the past.

At the present time, we are seeing the development of equipment and methods of using our traditional fuels more efficiently and also the development of new sources of energy for the production of heat. Heating technologists today should understand the traditional fuels and their use and also keep aware of new developments that are made in the heating industry.

2.5 TERMS TO KNOW

fuel
natural gas
combustion
pressure regulator
water gauge (wg)
manometer
resistance
pressure drop
liquefied petroleum gas (LPG)
gasify
fuel oil
combustion air
distillation
kilowatt per hour
electrical energy
electricmagnetic generation
solar heat
passive system
active system
geothermal heat
heat pump
coefficient of performance (COP)
heating seasonal performance factor (HSPF)
wood heating

3.1 INTRODUCTION

During combustion, chemicals in a fuel combine with oxygen from the air. In the process, new chemical compounds are formed, and heat is given off. We are interested in the heat, but analysis of the process also is important. This chapter covers the amount of heat that is given off during combustion, the products that are formed during combustion, and the requirements that make the combustion process safe.

3.2 COMBUSTION PROCESS

Combustion, the process of burning, is the release of heat energy from fuel through the process of oxidation. Combustion takes place when chemicals in the fuel combine with oxygen. The combustion process is the same process as the rusting of metal, but rusting is generally slow, and the heat produced is negligible. Burning a fuel is a relatively rapid process with a considerable amount of fuel and oxygen involved, so the amount of heat produced is greater. (See Figure 3-1). In order for combustion to take place, three elements must be present during the process. They are

1. Fuel
2. Oxygen
3. Heat

The **chemical elements** in the fuel represent a form of energy. Energy can be neither destroyed nor created, but it can be changed into other forms. This principle is the law of conservation of energy. In the combustion process, energy in the fuel is changed into heat and light, which are other forms of energy.

Fuels and Combustion

Fuels are composed of hydrogen atoms and carbon atoms in different combinations according to the particular fuel being considered. When a fuel goes through the combustion process, the atoms of carbon and hydrogen in the fuel com-

CHAPTER

3
Combustion

Chemical elements – Basic chemical components of a fuel.

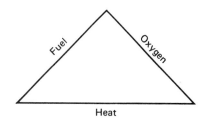

FIGURE 3–1
Combustion triangle representing the three necessary elements for combustion to take place

Kindling temperature – The temperature at which a fuel and oxygen mix will begin to burn.

bine with oxygen atoms from the air to form new elements, and heat is given off in the process. Methane (CH_4), which is the most common chemical compound found in gaseous fuels, has a chemical composition of four hydrogen atoms and one carbon atom. When methane combines with oxygen at a high enough temperature to start a reaction, the carbon atom combines with two atoms of oxygen to form carbon dioxide (CO_2), and two atoms of hydrogen combine with one atom of oxygen to form water (H_2O). The critical temperature is called the **kindling temperature.** Figure 3-2 shows how the chemicals in methane and oxygen combine in the combustion process and the products that result.

Fuels with a chemical composition different from methane produce slightly different results in their products of combustion, but since hydrogen and carbon are the main chemical components of all fuels, the results are similar.

Heating technicians do not need a complete understanding of the chemical processes involved in combustion, but they need a knowledge of the basic process, since flue gas analysis, which measures the chemicals in the products of combustion, is one method they use to measure the efficiency of the process.

Oxygen for Combustion

The oxygen for the combustion process is obtained from air that is mixed with the fuel, or combines with the fuel in

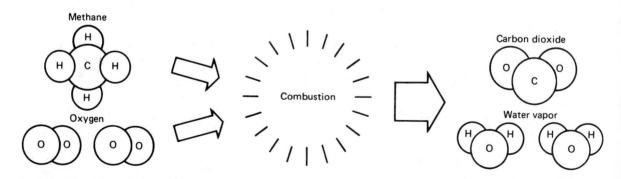

FIGURE 3–2

Combustion of methane and oxygen. In the combustion process, the elements in the fuel combine with oxygen to produce heat and other combinations of the same elements.

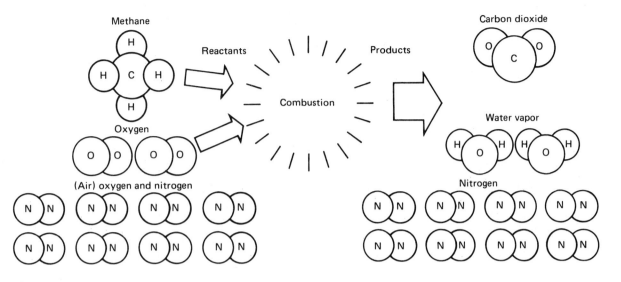

FIGURE 3-3 ———
In the combustion process air is used to supply oxygen. Air also contains nitrogen, but this passes through the process unchanged. The products of combustion contain the same amount of nitrogen as the air used in the process.

the flame, during the combustion process. Normal air, as we know it in our atmosphere, is made up of 76 percent nitrogen and 23 percent oxygen. There are traces of other elements in the air, but they do not affect the combustion. the nitrogen passes through the combustion process unchanged, and only the oxygen combines with the other elements in the fuel. The diagram in Figure 3-3 shows how the nitrogen passes through the process.

Heat for Combustion

In a typical burner, the fuel and air are mixed in the proper proportions to form a combustible mix. The final requirement for combustion to take place is that a kindling temperature be reached. The chemical reaction of the oxygen in the air and the hydrocarbons in the fuel will take the place only if the temperature is high enough to start the process. To start combustion, a spark or pilot flame is used. After the flame is established, the heat from the fire itself is great enough to keep the process going. If at any time the temperature of the flame is reduced below the kindling temperature, combustion ceases and the flame

Products of combustion – The chemical elements and compounds that are formed in the combustion process.

Complete combustion*

Incomplete combustion*

Impinge*

***Term defined in text**

goes out. The ignition devices that generate the heat for kindling the fire are discussed in Chapters 7, 8, and 11.

3.3 PRODUCTS OF COMBUSTION

Whenever a fuel is burned completely, the **products of combustion** are harmless. But if not enough air is present during the combustion process, some of the resulting products will be harmful.

Complete Combustion

When sufficient air is present during the combustion process, the chemical reaction is such that the carbon and the hydrogen in the fuel combine completely with the oxygen in the air. This process is called **complete combustion.** The new compounds, or combinations of elements formed, are products that are harmless to humans and the atmosphere. The main products are water (H_2O) and carbon dioxide (CO_2). Figure 3-4 shows these products of complete combustion in relation to the fuel and air.

Incomplete Combustion

When not enough air is present during the combustion process to react with the hydrogen and carbon completely, products that are harmful to people are formed. This process is called **incomplete combustion.** During the process, some of the oxygen and the hydrogen still combine, there is a shortage of oxygen. The resulting compound formed is carbon monoxide (CO) instead of carbon dioxide (CO_2). Carbon dioxide is a harmless gas; carbon monoxide is a deadly poisonous gas. Other products of incomplete combustion are aldehydes and soot. Aldehydes are toxic chemical compounds that are irritating to the eyes and nose. Soot is unburned carbon. Figure 3-5 shows the products of incomplete combustion in relation to the fuel and air.

Incomplete combustion also occurs when a flame is cooled below its ignition temperature—for example, when the flame **impinges,** or strikes, a cool metal surface. Unburned particles of carbon in the flame stick to the metal surface as soot. When soot is observed in a flame,

FIGURE 3–4 _____
Products of complete combustion, none of which are harmful to people

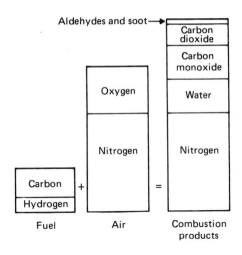

FIGURE 3-5 _____
Products of incomplete combustion, some of which are harmful to people

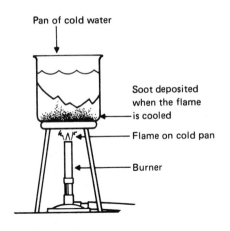

FIGURE 3-6 _____
Incomplete combustion. Soot, or unburned particles of carbon, is deposited on pan bottom when flame is cooled by pan.

incomplete combustion is taking place. Figure 3-6 illustrates how soot is formed on a pan bottom. Incomplete combustion can occur inside a furnace unit.

3.4 COMBUSTION AIR

Burners in heating equipment are designed to mix air and fuel in the proper proportions for complete combustion. The designer of the heating system must make sure that a sufficient amount of air can get to the equipment for combustion; the heating technician who sets up the unit must adjust the burner for the proper amount of air.

Ignition Limits

Combustion can take place only when the **air-fuel mix** is "right." A rather precise amount of oxygen per part of fuel is needed to form the combustion mix. Figure 3-7 shows that for natural gas, the mix has to be between 4 percent and 14 percent gas, and the remainder air, or combustion will not take place. Less than 4 percent gas is too lean a mixture, and more than 14 percent gas is too rich. Each

Air-fuel mix – The mixture of air, to provide oxygen, with a fuel to make a combustible combination.

FIGURE 3–7 _____
Percentage requirements of air-fuel mix for
combustion of natural gas

FIGURE 3–8 _____
Combustion air requirement for complete combustion of natural gas

fuel has its critical mix proportions. Combustion will not
take place unless the mix is right for the particular fuel
being used.

Air Quantities Required

Each fuel requires a certain proportion of air to the amount
of fuel being burned. If not enough air is available, dangerous products of combustion are formed. Figure 3-8 shows
the amount of air required for proper combustion for natural
gas.

Since the air-fuel mix is so critical, extra air must be provided to the burners. This air is called **excess air.** A system
should always be designed to provide at least 50 percent
excess air, or half again as much air as is actually needed for
complete combustion.

Excess air*

3.5 FLAME TYPES

Two types of flame are created by the combustion processes: blue and yellow. Each type is characteristic of the way the air is mixed with the fuel in the firing process. The burners used in the heating unit provide the air and fuel mixing, so ultimately the flame type is related to the type of burner used and to the type of fuel burned.

With some burners and fuels, a large proportion of the combustion air can be mixed with the fuel before the fuel is burned, and the blue flame results. An example is a burner that mixes air with **gaseous fuels,** such as natural or manufactured gases. In a burner for **liquid fuel,** such as fuel oil, or solid fuel, such as coal, the air cannot be mixed with the fuel before burning. In this case, the flame generated is yellow.

Blue Flame

If the air and the fuel are premixed (that is, before being introduced at the burner ports), the flame produced will be blue. This flame is very clean. This flame is called a Bunsen flame, after the burners used in laboratory work. In Figure 3-9, a cutaway view of a Bunsen burner shows the fuel being introduced through an orifice, or opening, in the burner body, or tube, and air is also introduced at this point. The fuel and air then mix as they travel up the burner body. When they are fired at the burner port, they produce a clean-burning blue flame. The blue flame is produced by the rapid burning of the hydrogen and the carbon in the fuel.

Gaseous fuels are easily mixed with air before the combustion takes place. The flames they produce usually are blue.

Yellow Flame

If the air cannot be mixed with the fuel before the fuel reaches the burner head, or if insufficient air is mixed with the fuel before the fuel is fired, the flame will be yellow. The yellow in the flame is caused by the carbon particles that are not completely burned. They exist inside the mantle of the flame and become incandescent from the heat. They do finally burn, but not until they mix with air that is drawn in from the sides of the flame.

Gaseous fuels – Fuel that is in the gaseous state when introduced into the burner.

Liquid fuel – Fuels that are introduced into the burner in the liquid state and vaporize only as they are burned.

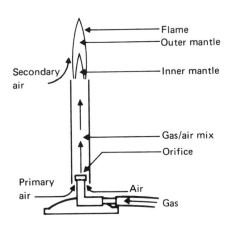

FIGURE 3–9 ———
Cutaway view of a Bunsen burner

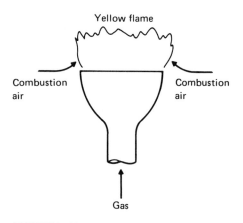

FIGURE 3–10
Yellow flame burner, in which the combustion air mixes with the fuel in the flame itself

Yellow, or incandescent, flame burners were used in nineteenth-century gaslights for lighting purposes. Because of the incandescence, the yellow flame burner gives off a great deal of light but less heat. The light is an unwanted form of energy for our purposes. See Figure 3-10 for a typical yellow flame burner.

Liquid fuels will not burn in the liquid state. They must be vaporized, and then the vapors mixed with air before they will burn. Figure 3-11 shows a simple pot burner with oil vapor and air burning. Fuel oil, a liquid, is vaporized and mixed with air as it leaves the burner face.

A fuel oil burner called a power burner is often used to achieve a more efficient mix of air and fuel oil. In a power burner the oil is pressurized by a fuel oil pump, and then flows to a nozzle in the burner face. The oil is atomized as it flows through the nozzle, and sprays into a combustion chamber in a definite pattern. As the atomized oil flows into the combustion chamber it is combined with combustion air that is blown through the burner face also. The air and oil mix to form a combustible mix. An oil burner face is shown in Figure 3-12.

Because of the difficulty of mixing the vaporized or atomized fuel and the air, a yellow flame is normally produced with an oil burner. Oil burners have been developed that work almost as efficiently as blue flame burners, but unless the fuel and air can be mixed thoroughly before igni-

FIGURE 3–11
Simple pot burner with oil vapor and air burning. Vapor burns after air is mixed above screen and vaporization is caused by heat from flame

FIGURE 3–12
Diagram of an oil burner face showing how the oil is vaporized and then mixed with air to make a combustible mix

tion, it is difficult to produce as efficient a flame with an oil burner as it is with a gas fuel burner.

Solid fuels (such as coke, wood, and coal), can be burned only as vapors. The heat from the process must vaporize the fuel, and then the vapors must mix with air for combustion to take place. The type of flame produced depends on the amount of air that can be mixed with the vapors. The amount of air is determined by the design of the burner and the burner system. Solid fuel burners are not covered in detail in this text.

3.6 COMBUSTION EFFICIENCY

The elements in the air and the fuel always combine in exact proportions in the combustion process. The elements in the combustion air react with the elements in the fuel so that the elements in the flue gas, while they have combined into different compounds, still equal in volume the elements in the air and the fuel. The bar graph in Figure 3-13 illustrates this principle.

By analyzing the elements in the fuel and those in the air, a technician can predict what elements will appear in

FIGURE 3–13

Graph showing how the products of combustion always have the same volume as the fuel and the air used in the combustion process

TABLE 3–1
Flue Gas Analysis for Natural Gas Combustion
Process, By Volume

Element	Percent of total volume
CO_2	8.86
O_2	10.70
CO	0.00
N_2	80.44
Total	100.00%

the flue gas, the efficiency of the process can be determined.

In a combustion process operating at maximum efficiency, the oxygen in the air, approximately 21 percent, combines with the carbon and hydrogen in the fuel and forms carbon dioxide and water vapor. The nitrogen in the air goes through the process unchanged. The products of combustion have carbon dioxide and nitrogen in the fixed proportion depending upon the amount of carbon and hydrogen in the fuel and the amount of combustion air used.

For natural gas or fuel oil, the percentage of carbon dioxide in the products of combustion is around 8 percent to 10 percent. The maximum possible is 12 percent. Table 3-1 shows the results of an analysis for a natural gas combustion process.

Flue Gas Analysis

Flue gas analysis is used to check the operating efficiency of a burner in a furnace or a boiler. The results of an analysis indicate if the equipment is in good condition and/or if it needs adjustment or replacement.

Three tests are required for making a simple analysis: (1) a CO_2 analysis, (2) a smoke spot test, and (3) a stack temperature reading. If a manual test method is used the tests are made by sampling the flue gas with the appropriate instruments and by measuring the temperature rise through the furnace. Most combustion tests are made either by using electronic test instruments and taking readings manually, or by using such instruments that are permanently installed on a unit, and indicate data continuously.

Instruments Used for Analysis

A CO_2 analyzer is a device that gives the CO_2 content of the flue gas leaving an operating unit in percentages. An analyzer used for manual testing is either a tubular instrument containing a liquid that will absorb CO_2, or a hand held instrument. (See Figure 3-14).

To use a CO_2 analyzer, a technician draws flue gas out of a furnace vent where flue gas leaves an operating heating unit. In a manual system the flue gas mixes with the liquid in the analyzer. In an electronic intstrument the flue

FIGURE 3–14
CO_2 analyzer being used by a technician for a combustion test

gas is analyzed as it goes through an ionizing cell. The analyzer indicates the amount of CO_2 present in the flue gas in percentages.

A smoke spot tester shows whether combustion is complete by indicating the presence of smoke in the flue gas. Smoke is an indication that all of the carbon in the flue is not burned in the combustion process. A smoke spot tester to be used manually is a vacuum pump with a filter paper holder located where the air that is pumped out of the flue passes through the paper. An electronic smoke spot tester also analyzes a sample of smoke drawn out of the flue of an operating unit, and results are shown on the instrument itself. Figure 3-15 shows a typical smoke spot tester.

To use a manual smoke spot tester, the technician inserts a flexible tube connected to the end of the tester pump into the flue, and a sample of flue gas is drawn out by the pump. The sample is drawn through a piece of filter paper by the pump. Stain on the paper indicates the presence of smoke in the flue gas. The stain on the paper is compared with a set of test stains to indicate the amount of smoke. Figure 3-16 shows a sample of the filter paper and the comparison chart used for grading the sample.

Stack temperature is measured with a thermometer that reads high temperatures. The stack temperature is then compared with the room temperature to show the temperature rise through the unit.

Procedure

During a flue gas analysis, the burner of a unit to be tested is first adjusted so it is operating correctly. The heating unit is run for 10 minutes so it will stabilize, and then the CO_2 sample is taken from the flue below the point at which any air enters—that is, below the draft diverter or the barometric damper. A smoke spot test is made at the same time from the same place. The smoke spot test should indicate only a trace of smoke, and the CO_2 reading should be between 8 percent and 10 percent. If the smoke is more than a trace, then more combustion air should be admitted at the burner. The CO_2 should be rechecked to make sure it does not go below 8 percent. If the smoke cannot be kept to a trace, the burner should be checked to ensure that all of the parts are operating correctly, and the

FIGURE 3–15
Smoke spot tester

FIGURE 3–16
Filter paper and smoke spot comparison chart

FIGURE 3–17 _____
Stack temperature thermometer being used on a stack for checking gas temperature

FIGURE 3–18 _____
Combustion efficiency slide rule showing the various tongues used for different fuels

furnace should be checked for air leaks into the combustion chamber.

To complete the analysis, the flue gas temperature is measured. The procedure is to insert a high-temperature thermometer in the flue at the same place where the smoke spot and CO_2 reading were taken. See Figure 3-17 for an illustration of a typical flue gas thermometer.

The flue gas temperature should be compared to the room temperature. The difference between the two, or flue gas temperature minus room temperature, is the net stack temperature.

Percentage Efficiency

A slide rule or graph supplied with the CO_2 analyzer is used to find the operating efficiency of the unit after the tests are made. A typical slide rule is illustrated in Figure 3-18.

The slide rule is made up of a body and interchangeable tongues. Each tongue is calibrated for a specific fuel. Make sure the tongue used is the one calibrated for the fuel used in the unit tested. The tongue is set so the net stack temperature, or temperature rise through the unit, appears in the window on the upper right. Find the CO_2 percentage, which is read on the analyzer in the window in the right center of the slide rule, and the combustion efficiency in percentage will be shown on the rule opposite the CO_2 figure.

The combustion efficiency should be 75 percent or more for a unit in reasonably good condition. If a unit cannot be adjusted to give at least 75 percent efficiency, it should be reconditioned or replaced.

3.7 SUMMARY

The controlled production of heat is necessary to maintain comfortable conditions in homes and business buildings. It also is important in the production of materials and equipment. The control of heat requires an understanding of the combustion process.

Fuel, oxygen, and a high enough temperature to start a reaction are required for the combustion process. The use of different fuels and the amount of fuel used will determine how much heat is produced in a given time. The method used to introduce the air, which provides the oxygen, into the process determines the type of flame produced. The application of the heat to start the process is a function of the burner.

Under the proper firing conditions, the combustion process is safe and will not produce dangerous products. If insufficient air is furnished to the flame, however, by-products can be produced that are extremely hazardous to people and animals. A designer must always be sure that a burner has not only enough air for proper combustion but excess air, to prevent the development of a dangerous situation.

3.8 TERMS TO KNOW

combustion process
chemical elements
kindling temperature
products of combustion
complete combustion
incomplete combustion
impinge
air-fuel mix
excess air
gaseous fuels
liquid fuel
blue flame
yellow flame

CHAPTER

4

Heating Mediums

Warm air out

Flue outlets

Metal clamshells

Burner openings

Cool air in

FIGURE 4–1
Warm air heat exchanger

Medium – A gas or liquid used to carry heat from one place to another in a heating system.

Heat exchanger*.

*Term defined in text.

4.1 INTRODUCTION

In a typical heating application, it often is necessary to generate heat at one point in a building, while heating rooms or spaces that are remote from that point. For example, the furnace in a large building often is located in an equipment room in the basement of the building, but this furnace is used to heat the entire building. Because of the difference in location, some **medium** must be used to carry the heat to the spaces where the heat is needed. Air, water, and steam are the most common mediums used for this purpose. Each of these mediums is discussed in this chapter, with emphasis on air and water.

4.2 HEAT EXCHANGE

A part of the furnace or boiler called a **heat exchanger** is used to extract heat from the fire and products of combustion and to heat the air or water that is then used to heat the building. The actual design of heat exchangers is covered later in this text, but a brief description is given here. A heat exchanger is any device that extracts heat from one medium and transfers it to another. Figure 4-1 illustrates the type of heat exchanger used in hot air furnaces.

In the case of a hot air furnace, the exchange is made between the hot products of combustion on the inside of the exchanger and the air that passes over the outside. In a hot water or steam boiler, the exchange is made between the hot products of combustion on the one side of the boiler tubes and the water that is on the other side of the tube walls.

Rate of Heat Exchange

The formula in Chapter 1 relating to the amount of heat involved when a temperature difference occurs in any given material will help explain what happens when heat is transferred through the walls of a heat exchanger. In that formula,

$$Btu = ms(T_1 - T_2)$$

But when heat passes through any material, because of a

temperature difference between two mediums on each side of the material, the formula must be revised to account for the movement of the mediums, the thin film of air or water on each side of the material, and the fact that the heat is removed on one side as fast as it is made available on the other. Therefore, it is necessary to use a factor called a **heat transfer factor,** or **conduction factor,** in the calculations. A heat transfer factor, normally called a U factor, is one that has been calculated by considering the variables. The heat transfer factor actually is a number that defines the amount of heat that will pass through a given material in one hour per square foot of material, per 1°F temperature difference on each side of the material. The formula for heat transfer by conduction is

$$Btu/h = A \times U \times (T_2 - T_2)$$

where

Btu/h	= amount of heat transferred in Btu per hour
A	= area of material to be considered
U	= conduction factor for given material
$(T_1 - T_2)$	= temperature difference between the two sides of the material

In practical applications, technicians do not need to calculate the heat output for the heat exchanger used. The manufacturer of the equipment furnishes the Btu/h output for the heat exchanger as part of the catalog data for the heating unit.

4.3 AIR AS THE HEATING MEDIUM

In a warm air system, the furnace has a burner in which combustion takes place, and a heat exchanger where most of the heat is extracted from the flames and the products of combustion. Figure 4-2 provides a cutaway view of a furnace showing the heat exchanger in place.

The heat that is extracted by the heat exchanger from the flue gases is added to the air distributed through the building for heating. In the usual design of exchangers, there is only a thin metal separation between the hot products of combustion and the circulating air, so the rate of exchange is very high; 80 percent is the usual efficiency of a warm air exchanger.

Heat transfer factor (conduction factor)*

Downflow furnace

Blower

Heat exchanger

FIGURE 4–2
A warm air heat exchanger and a cutaway view of a downflow furnace with the heat exchanger in place

Temperature rise (TR)*

Air being used as a heating medium is blown either directly, or through supply ducts, to the building spaces requiring heat. The circulation of the air is accomplished by a blower, or air mover. Hence the term forced air is used to describe this type of system.

Provision must be made for return air to get back to the heating unit, to ensure complete circulation. As warm air circulates through the spaces requiring heat, the air gives up its heat. The air that is now at normal room temperature is returned to the furnace, to start the cycle over.

Temperature Rise of the Air

Temperature rise through a furnace is the difference between the temperature of the air entering the furnace and the air leaving. The term thus indicates the amount of heat the air has picked up as it passed through the heat exchanger. Temperature rise is a function of the amount of heat available, the efficiency of the exchange, and the amount of air going through the exchanger.

Each type of furnace is designed to work best at some particular temperature rise. Combustion-type furnaces should have from 80°F to 100°F temperature rise, electrical furnaces less than 50°F, and heat pumps only about 20°F. It is important to check the temperature rise through a furnace since a high temperature rise can indicate that the heat exchanger may be too hot and can be damaged. A low temperature rise indicates that too much air may be circulating for comfort.

Temperature rise can be determined by measuring the entering and leaving temperatures with a thermometer and then subtracting the entering temperature from the leaving temperature, as in the following formula:

$$TR = T_L - T_E$$

where:

TR = temperature rise
T_L = temperature leaving
T_E = temperature entering

This direct approach is a good one to use on an existing system when access to the unit is easy.

When a distribution system is designed, the quantity of air to circulate is often decided upon for a building of a given size, independently of the size of furnace used. In this case, the temperature rise through the furnace, should be checked to make sure the air quantity is correct for the heat exchanger. This check can be done by using the Btu/h output of the furnace and the measured temperature rise through the furnace. The formula to do this is as follows:

$$TR = \frac{Btu/h}{cfm \times 1.08}$$

where:

TR	= temperature rise
cfm	= quantity of air in cu ft per minute
1.08	= constant, based on specific heat of air 60 min in 1 hr

This formula is for air **standard conditions.** Standard conditions refers to temperatures and relative humidity condition considered standard. In most heating-only applications, it is not necessary to be concerned with moisture changes in the air as the temperature changes, so the temperatures in the formula are for dry bulb or **sensible readings** only.

To use the formula, take the rated heating output of the furnace in Btu/h and divide that by the quantity of air in cfm going through it, times the constant, 1.08. The result is the temperature rise of the air as the air passes through the exchanger.

EXAMPLE 4-1

If a furnace is rated at 64,000 Btu/h output, and the blower is adjusted to deliver 960 cfm of air, what is the temperature rise of the air through the furnace?

Solution

$$TR = \frac{Btu/h}{cfm \times 1.08}$$

$$= \frac{64,000 \; Btu/h}{960 \; cfm \times 1.08}$$

$$= 61.7°F$$

Sensible reading – A temperature that can be sensed by a human being or measured by a standard thermometer.

Hydronic – A heating system that uses a liquid as the primary heating medium.

Terminal device*

FIGURE 4–3 _____

Typical small boiler showing the circulating pump and burner location

The answer is 61.7°F. The temperature of the air leaving the furnace will be 61.7°F higher than the temperature of the air entering the furnace.

Note: The figure 1.08 is derived by using values related to standard air. Recently the value of 1.1 has often been assigned to this figure with the knowledge that air standards vary. This is more often the practice in air conditioning applications rather than in heating applications. In this text, the value of 1.08 will be shown, but results found using 1.1 will not vary enough to be a problem.

4.4 WATER AS THE HEATING MEDIUM

In a hot water, or **hydronic,** heating system, the circulating medium that carries the heat to the point of use is water. The burner is located under a boiler, and water is heated by the flames and products of combustion. A picture of a small gas-fired boiler is shown in Figure 4-3. A number of different types of boilers are used, but they all have the same basic purpose: heating water.

The hot water is circulated through a piping system to the various rooms or zones in which heat is needed. Within the zone the water is directed through a **terminal device,** such as a radiator or a coil. In the terminal device heat from the water is used to warm air. The warm air in turn is used to heat the zone or space. The system uses a supply and a return pipe, as well as a pump or pumps, to circulate the water. Water is an excellent medium to carry heat because of its high specific heat capacity. But since the ultimate purpose is to heat air, hydronic systems are most practical when the heat is to be utilized at some distance from the point of generation.

To understand why a hydronic system transfers heat so well, we must look at the specific heat, or the amount of heat the water will hold per unit. Water is used as the basis of comparison for the specific heat of all substances, and the specific heat of water is considered 1.0. Thus, 1 Btu of heat is required to raise the temperature of 1 lb of water 1°F. The value 1.0 is relatively high compared to that of other materials listed. Table 4-1 shows the specific heat of some common materials, including water. Where the specific heat of water is 1.0, the specific heat of air is 0.24, which shows that water will carry approximately four times as much heat as air, per given weight.

Temperature Rise of Water

It is sometimes necessary to calculate the temperature rise of water through a boiler, which is only a heat exchanger, when designing a hydronic system. The amount of heat carried by the water is a function of the temperature of the water, and the quantity of water in gallons per minute (gpm), that is circulating.

Since water is used as the standard for defining specific heat, and we know that 1 Btu is required to raise the temperature of 1 lb of water 1°F, a formula can be derived using these data to find the temperature rise of water when water is heated.

The usual way of measuring the rate of flow of water in a system is by gpm, and since specific heat is calculated per pound of water, the formula has to convert pounds of water to gallons and hours to minutes. The formula is

$$TR = \frac{Btu/h}{gpm \times 500}$$

where:

TR = temperature rise of the water flowing through the boiler
Btu/h = total heat output rating for the boiler
gpm = quantity of water passing through the boiler in gpm
500 = a constant derived by multiplying the specific heat of water
× the minutes in an hour
× the weight of water in pounds per gallon

To figure the temperature rise of water through a given boiler, divide the rated Btu/h capacity of the boiler by the amount of water flowing through it in gpm, multiply by 500. The answer will be the temperature rise of the water in degrees Fahrenheit.

TABLE 4-1
Specific Heat of Some Common Materials

Substance	Specific heat (Btu/lb/°F)
Air	0.24
Aluminum	0.22
Copper	0.093
Glass	0.21
Ice	0.50
Iron (steel)	0.115
Silver	0.056
Steam	0.48
Water	1.00

Condensate – A liquid obtained by condensation of a gas or vapor.

EXAMPLE 4-2

If a boiler with a rated output capacity of 450,000 Btu/h is installed on a system in which water is circulated at a rate of 36 gpm, what will be the temperature rise of the water?

Solution

$$TR = \frac{Btu/h}{gpm \times 500}$$

$$= \frac{450,000\ Btu/h}{36gpm \times 500}$$

$$= 25°F$$

The temperature of the water leaving the boiler will be 25°F higher than the entering temperature.

4.5 STEAM AS A HEATING MEDIUM

When steam is used as a medium to carry heat, the system is similar to that described for hot water. The main difference is found in the boiler and the piping for distribution and return of **condensate.** Since the heat from the burner is used to raise the temperature of water high enough to convert it to steam and since steam has so much more volume than the liquid, the system operates at pressures above atmospheric. The steam, under pressure, circulates to the terminal devices in the system, where heat is extracted from the steam by those devices. Heat extraction condenses the steam into water, and so the return system conducts water back to the boiler, to start the cycle over.

Steam is seldom used as a medium in small or mid-sized heating jobs. Steam tables giving the Btu content of steam per pound of steam at various pressures and temperatures are normally used when designing a steam system and should be referred to by the designer of any such system.

4.6 SUMMARY

The practical use of heat produced by the combustion process requires the distribution of that heat throughout the building or the spaces to be heated. Since the products of combustion themselves may be harmful to people in the building, the heat must be extracted from the products of combustion and distributed through the building by using some medium. The most common mediums used for this purpose are air, water, and steam.

The heat is extracted from the products of combustion by a heat exchanger. The flue gases and heating mediums are completely separated from each other in the heat exchanger. The rate of heat exchange between the flue gas and the medium can be calculated accurately, and the amount of heat carried by the medium can be predicted. The calculations enable designers to determine exactly how much air, water, or steam must be distributed to the different areas of a building to provide the required heat to those areas.

Each medium has particular characteristics that makes it use more practical for some applications than others. Since air is the medium finally used for heating the spaces occupied by people, air should be used as the primary medium whenever possible. If heat has to be transferred long distances, water or steam should be the primary medium and air should be the secondary medium. The best medium to use in each individual case is determined by the designer after analyzing all of the heating requirements of the building.

4.7 TERMS TO KNOW

medium
heat exchanger
conduction factor
heat transfer factor
temperature rise (TR)
standard conditions
sensible readings

hydronic
terminal device
condensate

5.1 INTRODUCTION

The two major types of systems used for heating buildings are the warm air system and the hydronic system. This chapter covers the warm air system, and Chapter 10 covers the hydronic system.

In a warm air system, air is heated in a furnace and circulated through a building to maintain a desired temperature in the building. Many different types of furnaces are used in this system. These furnaces are categorized by fuels, heating capacity, building use, physical arrangement, and specific application. This chapter describes the total system and the general types of furnaces.

5.2 WARM AIR HEATING SYSTEM

A warm air heating system is one in which the heating medium is air. This air is heated as it goes through a furnace. The heated air is then circulated to various areas in a building where heat is needed, through a system of supply ducts. The air is distributed into building spaces through supply registers, and is then returned to the furnace through return air grilles and return air ductwork. The air is circulated by a blower that is part of the furnace or system.

The temperature of the air in the building is controlled by a **thermostat** located in the building. A thermostat senses the temperature of the air and "brings on" the furnace if the temperature drops. When the temperature of the air rises, the furnace is "turned off."

Warm air systems are most practical in buildings where the furnace can be located relatively near the areas needing heat. Warm air ducts are usually fairly large and require a considerable amount of space in the building frame.

5.3 GENERAL TYPES OF FURNACES

Three main types of furnaces are available. The first is related to the fuel or energy used to produce the heat, the second is related to the type of job that the unit will be used on, and the third is related to physical characteristics

CHAPTER

5

Warm Air Systems and Heating Equipment

Thermostat*

* Terms defined in text.

of the unit itself. Each of these is discussed in detail in the following sections.

Furnaces for Different Fuels

Furnaces burn gaseous fuels, liquid fuels, and solid fuels; others use electric energy to produce heat. Each of these types of furnaces has some features in common, but each is also different from the others in some ways. Gas furnace burners and heat exchangers are different from those used for oil. Electric furnaces use resistance heaters instead of burners and heat exchangers. Details concerning gas, oil, and electric furnaces are given in the following chapters, but since furnaces that burn solid fuels are not common in smaller sizes, they will be covered only in a general way.

Residential and Commercial Furnaces

In the second category, units by size, units used for residences and smaller commercial jobs are generally smaller in heating capacity and also are more compact in physical size than are units built for commercial or industrial applications. Furnaces rated as residential type normally have an input rating up to 300,000 Btu/h, while commerically rated furnaces may be rated in millions of Btu/h. Control systems are less sophisticated on residential units than on commercial units. Some options, such as multiple fuel use, are not available on residential units but are common on commercial units. Commercial units are built with more concern about utility and less about appearance, and they may include safety control features important to commercial applications.

Physical Characteristics

Since nearly every application is different, heating units must be built to fit different types of installations. This concern brings us to the third category for furnaces: physical configuration. Furnaces in this category are identified by the direction the air passes through the unit. In this respect, furnaces are built to operate as **upflow, downflow,** or **horizontal furnaces.** These three configurations enable designers to find units that will fit almost any job.

Upflow Furnaces. A typical upflow furnace has provisions for the return air to enter near the bottom of the unit. The air comes through the bottom of the cabinet or on one side. The air is drawn through the filters by a blower, passes through the **heat exchanger,** and goes out at the top of the unit. Figure 5-1 shows a typical upflow gas-fired furnace.

When upflow furnaces are used, they usually are placed below the ductwork. The air can go up from the furnace and straight into the ducts. An example is a system used in a residence with a basement, where the ductwork is located on the ceiling of the basement.

Downflow Furnaces. In a downflow furnace also called a counterflow furnace), air enters the unit at the top, passes through the filters, is pulled through the blower, and is then blown down through the heat exchanger. The air is discharged vertically out of the bottom of the unit. Figure 5-2 shows a typical downflow gas-fired furnace.

Downflow furnaces are most practical in applications where the ductwork is located below the unit. The air can then go straight down from the unit into the ductwork. This type of system is often used when the floor is a concrete slab-on-grade and the ductwork is located under the slab. One of the advantages of using a downflow furnace in this type of installation is that the furnace can be located in a closet on the main floor of the building; therefore a special equipment room does not have to be provided under the floor.

Horizontal Furnaces. A horizontal furnace is one that has air pass through it horizontally. That is, the air enters horizontally on one end, passes through the furnace, and is discharged horizontally on the other end. The air enters the furnace, passes through the filter, is drawn through the blower, goes across the heat exchanger, and is discharged. This type of furnace is available in both left- and right-hand discharge, and the unit has to be ordered accordingly. Figure 5-3 shows a typical horizontal gas-fired furnace.

Horizontal furnaces are most often used when the ductwork is on the same level as the unit and when the height of the space of the furnace is to be installed in is limited. A good example is the use in a crawl space or in the attic of a building. In warmer climates where supply

Upflow furnace – A furnace in which the air to be heated enters through the bottom or one side and is discharged through the top.

Downflow furnace – A furnace in which the air to be heated enters at the top and is discharged down through the bottom.

Horizontal furnace – A furnace in which the airflow is horizontal (air to be heated enters one end and is discharged out the other).

Heat exchanger – A device that is used to transfer heat from the flue gases in a furnace to the air used to heat a building.

FIGURE 5–1 _____
Typical upflow gas-fired furnace

FIGURE 5–2 _____
Typical downflow gas-fired furnace

FIGURE 5–3 _____
Typical horizontal gas-fired fur-
nace

registers can be located overhead, a horizontal furnace often is used in the attic, and the ductwork is run in the attic. The supply registers are then dropped down through the ceiling into the spaces to be conditioned. In colder climates where the registers should be located on the floor, the horizontal furnace is placed in the crawl space for the best heating. The ducts run under the floor with the **boots** for the registers going up through the floor.

5.4 UNITS FOR SPECIAL APPLICATIONS

In addition to the usual type of heating units where the heat source, filters, blowers, and other parts are all contained in one cabinet, other types of units are used for special applications. These include duct heaters to be installed directly in the duct of a system, unit heaters that are self-contained and are not connected to any duct system, and radiant heaters that provide heat to a particular location but do not heat the air in the space.

Duct Heaters

A **duct heater** is a heating unit that normally is installed in some part of a duct system remote from the blower unit. A duct heater contains a burner or heating elements, a heat exchanger, and the controls for the burner. A cabinet encloses these parts. Duct connections on each end connect to a duct system. The duct heater is used for heating air, just like any other furnace. Air must go across the heat exchanger to prevent burn-out of the exchanger. The air supply comes from a blower located in a part of the system remote from the heating units's location. The illustration in Figure 5-4 shows the arrangement of parts in a duct heater.

Duct heaters are manufactured to burn or use any of the common fuels, or to use energy available to any heating unit. Duct heaters often are used to provide heat to one part of a building or one zone in a building rather than the entire building. These heaters often are used in branch ducts and can be controlled individually to provide **zone control.**

For each of the fuels or energy used, the burners,

Boot – A sheet metal box that connects a register or grille, or diffuser to the duct-work.

Duct heater – A heating unit that is installed directly in an air duct. (The air is circulated by a remote blower.)

Zone control – Temperature control in a building by individual zones as opposed to the entire building as a whole.

FIGURE 5-4 _____
Drawing of duct furnace with cutaway showing burners in relation to heat exchanger

Blower control – A device used in a heating unit to turn the blower on when the burner comes on.

Flow switch – A device that senses whether air is moving in a duct.

Unit heater*

Infrared rays – Invisible rays that have a longer wavelength than visible light but can be used to transfer energy from one place to another.

exchangers, elements, and other heat-producing components are similar to those found in self-contained units.

Controls for duct heaters vary slightly from those found in central station units in that the air mover is remote from the location of the heater, and so the **blower controls** are not located near the blower. It is common practice to provide **flow switches** to "prove," or provide evidence of, the flow of air in a duct before a duct heater heating unit can come on. The flow switch is used to eliminate the danger of fire or damage to the unit or building because of high temperature in the unit if no air is moving across the heat exchanger.

Unit Heaters

Another type of heating unit, **the unit heater,** is used when close control of the temperature in a space is not required. A unit heater is self-contained, with the burners or heating elements, heat exchanger, controls, and an air mover all in one cabinet. This type of heater is not intended for use with a duct system. The air mover draws air in through the back of the unit, blows it through the heat exchanger, and is discharged out the front through a louver on the face. Figure 5-5 shows the parts of a unit heater.

Unit heaters are used for heating large open spaces, such as warehouse or factory bays. Usually such heaters are suspended from the ceiling. The louver on the front of a unit heater directs the air leaving the unit to the occupied space. Unit heaters can be fired with gas or oil, or electricity, steam, or hot water may be used as the source of heat. Because the unit heater is not used with a duct system and does not have to move air against the friction in the ducts, a propeller fan is used as an air mover. The propeller fan can move the air efficiently in free-flow applications.

The components in the unit heater, such as burners, heat exchanger, and controls, are similar to those found in self-contained furnaces for central installation on duct systems.

FIGURE 5–5
Cutaway drawing of unit heater

Infrared Radiant Heaters

A third type of heating unit is used for specialized applications. It is the infrared radiant heater. This type of heater emits **infrared rays** that do not heat the air but do heat any

FIGURE 5–6
Infrared heater with radiant surface in back of reflector cabinet

objects or surfaces that they strike. The typical infrared heater has a heat source (such as a gas burner), that heats ceramic elements to a very high temperature, and the rays are then given off by the hot element. Figure 5-6 shows an infrared heater.

In a typical application, infrared heaters are suspended from the ceiling or roof of a building space, and the heating rays are directed downward. The rays are changed into heat energy only when they strike the floor, people, or other objects in the line of sight of the heater. The air around the objects is then heated to some extent by convection as the air circulates across the warm objects.

Infrared heaters are practical as a source of heat in any building where it is not important that the air in the space be heated, but where it is important to provide heat to the floor area or to people on the floor level.

5.5 SUMMARY

The purpose of any heating system is to provide a comfortably warm atmosphere in a building. The system achieves this end by controlling the temperature of the air in the building by means of a warm air furnace.

Several types of warm air heating units are available. A unit should be selected only after the designer considers the fuel to use, physical requirements, and the airflow patterns required. In addition to the typical self-contained furnaces available, there are specialized units that may be better for particular applications.

5.6 TERMS TO KNOW

thermostat
upflow furnace
horizontal furnace
heat exchanger
boot
duct heater
zone control
blower control
flow switch
unit heater
infrared rays

6.1 INTRODUCTION

Warm air furnaces are the most commonly used types of heating units for small and medium-sized buildings. These furnaces are manufactured in various styles and for use with different fuels. Regardless of the style of fuel used, each type of furnace has some components that are similar: cabinets, blowers, filters, burners, and heat exchangers.

Cabinets, blowers, and filters are covered in detail in this chapter, while burners and heat exchangers are covered in a more general way. Burners and heat exchangers are explained in more detail in the chapters on each system.

6.2 WARM AIR FURNACES

In a warm air heating system, a furnace provides warm air for fheating a building or spaces in a building. There are many types of furnaces. Each type is different from the others in some respects, but all of them have certain parts that are similar. These similar parts are described in this chapter.

A warm air furnace is made up of the following five major components:

1. Cabinet
2. Blower
3. Filter
4. Burners
5. Heat exchanger

In addition, each type of furnace has a set of controls that operate it. Figure 6-1 shows a furnace with these parts identified.

Cabinet

Most furnaces have a sheet metal cabinet that supports the other components and forms an enclosure to direct the air through the blower and across the heat exchanger. Figure 6-2 shows a furnace cabinet.

The cabinet is made of sheet metal, painted to protect the metal from corrosion, and in most cases, insulated on

FIGURE 6-1
Cutaway view of upflow oil-fired furnace showing major components

FIGURE 6–2
Modern furnace cabinet

the inside to provide quiet operation. In smaller units, the cabinet is rigid enough to act as a support for the other components. Larger furnaces usually require a frame to support these parts. Much consideration is given to the appearance of the cabinet for residential furnaces, but units for commercial or industrial applications are designed for utility rather than appearance.

The cabinet is built to enclose the components on all sides except the front. Openings are provided where the return and supply air duct connections are made. If the unit has options regarding the location of the return air, such as an upflow furnace usually has, then knock-outs are provided where the openings can be cut. Knock-outs are indentations in the sheet metal that show where openings can be made.

The front of the cabinet usually is covered by a removable door or panel. This door encloses a vestibule that holds the burner and the controls. The heat exchanger is behind another panel in the back of the vestibule, and the burner, or burners, extends through this panel into the space below the heat exchanger. The front panel, or door, is slotted or louvered so combustion air can get into the burner vestibule.

The filters are located in a separate compartment with the blower. This compartment has an access door that can be removed for changing filters or for servicing the blower.

Blower

Most heating units for use in small- to medium-sized buildings are designed as package units. The blower, filter, burners, and controls are in one cabinet. The blower, or air mover, used in these units is almost always a centrifugal blower. Such blowers are designed to move the air efficiently in the duct system and with relatively low power consumption. The blowers can also be adjusted to deliver different air quantities.

Medium- to high-pressure air distribution systems are used in some larger buildings. The higher pressure enables the distribution of large quantities of air through smaller ducts. In these systems, axial-flow fans are often used. The axial-flow fan can develop higher pressure in the duct system than the centrifugal fan, but it uses more power in the process.

FIGURE 6–3 ————————————
Wheel used in centrifugal blower showing the
vanes connected to the rim

FIGURE 6–4 ————————————
Centrifugal blower scroll

Centrifugal Blower. The most commonly used blower
for a furnace is a centrifugal blower. This type of blower
uses a wheel that looks like a paddle wheel with small
blades to move the air. These vanes, as they are called, are
located around the outside of the wheel. A centrifugal
blower wheel is shown in Figure 6-3.

The wheel rotates on a shaft parallel with the vanes and
at right angles to the direction of air flow. Air is pulled in
through the middle of the wheel and is thrown out through
the vanes as they spin. The air is discharged through an outlet at one point on a sheet metal scroll surrounding the
wheel. Figure 6-4 shows a blower scroll without the wheel.

Most smaller furnaces in residential applications use a
direct-drive blower-motor combination, as shown in
Figure 6-5. In a direct-drive blower, the wheel hub of the
blower is mounted directly on to the shaft of the motor.
The wheel speed is the same as the motor speed.
Normally, a multispeed motor is used on a direct-drive
blower application, so control can be maintained on the air
output, which is directly related to wheel speed.

On some furnaces and most larger units, the blower is
belt driven. In this application, the motor is located adjacent to the blower, and the drive is through the pulleys
and a belt. Figure 6-6 shows a furnace with a belt-drive
blower. Such an arrangement makes it possible to change

FIGURE 6–5 ————————————
Cutaway view of upflow oil furnace with direct-
drive blower

Belt-drive blower assembly

FIGURE 6–6 _____
Belt-drive blower in modern furnace

FIGURE 6–7 _____
Axial-flow blower used for medium-and
high-pressure distribution systems

pulley sizes to provide blower speed adjustment for controlling air output. Most belt-driven systems have an adjustable pulley on the motor. This pulley allows for changing the blower speed, which changes the air quantity through a set range. The motor is mounted in such a way that belt tension can be adjusted as the pulley size is changed.

Axial-Flow Blower. In applications where air has to be provided at medium to high pressure, an axial-flow blower is used. This blower has a wheel that looks like a turbine wheel. The blades are at right angles to the shaft on which it turns and are inclined so they move air like a propeller does. The shaft itself is parallel to the direction of the airflow. The wheel is enclosed within a cabinet, which is normally part of the ductwork. An axial-flow blower is shown in Figure 6-7.

The motor that drives the axial-flow blower may be within the cabinet or adjacent to it. An axial-flow blower usually is found on duct systems on larger jobs.

Filters

Micron – A unit of measurement; 0.001 millimeter, or 1/25,400 of an inch

One of the advantages of a circulated air system is that all of the air in the building can be constantly filtered and cleaned. Air filters in the heating unit or in the return air duct system ahead of the unit perform this operation. In some cases, filters may be placed in the return air grilles where the air is taken out of each room. As the air in the building passes through the return air system, the air is cleaned as it goes through the filters.

Of the many different types of filters, the most common ones fall into one of two major categories. The first is mechanical filtration. It is represented by mesh-type filters that clan the air by **impingement** or by straining. The second category is electronic filtration. In this type, unwanted particles of dust and dirt are removed from the air by electronic means.

Filter Efficiency. The most important variations found in filters relate to the cleaning efficiency of the filter. Standard disposable filters furnished with most residential furnaces have a fairly low efficiency, as compared to filters used for commercial applications. The mesh of standard filters is efficient in filtering out large particles of dust and dirt but not small particles.

To rate filter efficiency, it is necessary to define particle size as related to the dust and dirt that must be removed from the air. The table shown in Figure 6-8 shows the approximate particle size of some common materials.

The particle sizes in Figure 6-8 are given in **microns.** One micron equals 0.001 millimeter (mm), or 1/25,400 inch. A typical sand particle is approximately 100 microns across; a particle of water in airborne fog is 10 microns in diameter. An optical microscope can be used to magnify particles down to 0.1 micron in size, but an electron microscope is required for smaller particles.

Mesh-type filters are relatively efficient in removing particles from the air down to somewhere between 1 and 10 microns, and an electronic filter will remove particles as small as 0.03 micron. The actual efficiency of filters is measured in the laboratory by using dust made up of particles of known size and then measuring the percentage of the dust removed from the air under a given set of conditions.

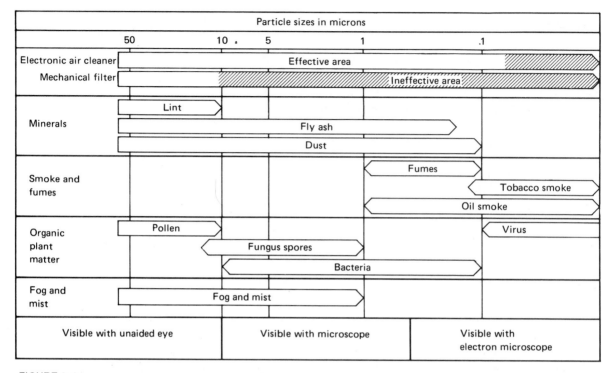

FIGURE 6–8

Size of some airborne particles in microns

Mechanical Filters. Two different principles of filtration are involved in mechanical filters: impingement and straining. In a filter that uses the impingement principle, the dust and dirt are collected on the filter media by sticking, or impinging, on a sticky surface. The surface is made sticky by coating the media with oil. In the straining filter, the fibers of the media are close enough together so the particles are simply caught between the strands. Usually the impingement filter is more efficient. Figure 6-8a illustrates some of the types of filters available.

Many different materials are used for media in mechanical filters, including fiberglass, plastic foam, shredded paper, and metal mesh. The media usually is formed into a pad or blanket and is used in that form. It may be enclosed in a cardboard frame and the frame then used in the filtering unit. If the pad is used alone, it is normally installed in frames that are part of the unit.

The performance of a mechanical filter in removing contaminants from the air is dependent upon the air going through the filter at a slow enough velocity and in such a path that it has to make rapid changes in direction. As the air goes through the filter, the velocity of the contaminants will cause them to continue in a straight line, even as the air changes direction, and they will impinge on the filter material and strike there. The efficiency of a filter can be increased by increasing the depth, thus making it necessary for the air to go through more changes of direction and consequently making it more likely that the dirt and dust in the air will impinge on the media.

Mechanical filters of the straining or impingement type can be used to filter particles from the air from about 100 microns down to 1 micron in size. By making the filters thicker and using material that is denser, the efficiency can be improved to where particles as small as 0.01 micron can be removed. For particles smaller than this, deep bed filters of special design are required. A deep bed filter is thicker and has more depth of filtering material. A deep bed filter for medium-high filtering efficiency is shown in Figure 6-9.

To select the proper filter, the designer must first decide on the percentage of efficiency desired. Then he or she will select a type and size of filter from a manufacturer's catalog that will provide that efficiency with the amount of air that has to be filtered. The size of the filter must be checked to ensure that it will fit the unit; and the **pressure drop** through the filters must be checked against the allowable pressure drop for the system so the blower can deliver the proper quantity of air.

Electronic Filters. Electronic filters are used in both residential and commercial applications when a high degree of filtration efficiency is desired. Figure 6-10 shows a typical electronic air cleaner as used for residential applications.

The electronic filter does not use filter media except as a prefilter. The actual cleaning is accomplished by passing the air from the system through a grid of wires that have been electrically charged with high-voltage direct current electricity. This prefiltering **ionizes** the dust particles in the air. The air, with the dust particles, then passes through a series of plates that are also electrically charged, and the ionized dust is attracted to and adheres to the

FIGURE 6–8A
Filters are available in many different sizes and types

Pressure drop – The difference between the air pressure on either side of a filter or other restriction in a duct system

Ionize – Create an electrostatic charge on particles by displacing some of the free electrons on the atoms of the particles

FIGURE 6–9 _____
Deep bed filter used for medium-high efficiency in mechanical filtration

FIGURE 6–10 _____
Electronic air cleaner used for residential or small commercial applications

plates by electrostatic attraction. Not only is the general efficiency of the electronic filter high, but the air cleaner will remove extremely small particles from the air. Figure 6-11 provides information on the general efficiency of the electronic filter.

An electronic filter is sized by using manufacturers data sheets as to its efficiency. In the sheets, the efficiency is shown for various cfm through the filter. The designer selects the filter that will give the desired efficiency at the quantity of air required for the job.

Special applications, such as rooms that require high-efficiency filtration, normally use filters that have great density and large area, and, in many cases, they are very

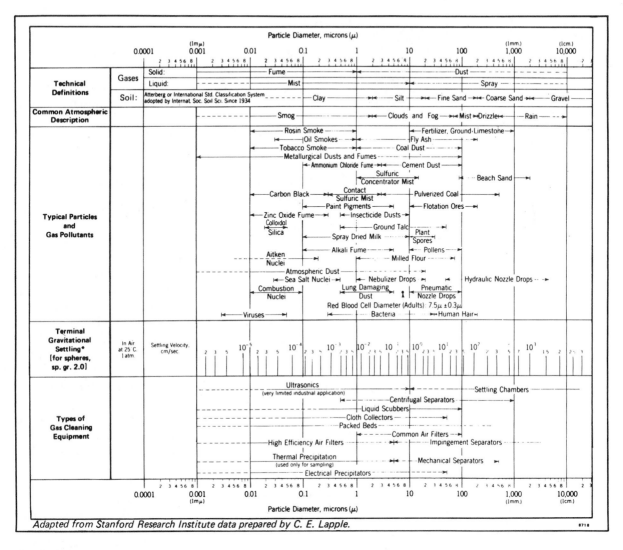

FIGURE 6–11
Size and characteristics of some airborne particles and filter efficiency related to them

deep. The resistance to airflow is high in such systems and is allowed for in the design of the total air system. A bag filter used for high efficiency cleaning is shown in Figure 6-12.

In some cases, filters are used for odor removal or control. Special chemical filters, such as charcoal filter beds, are used for this purpose.

FIGURE 6–12 _____
Bag filter used for high efficiency and long life in mechanical filtration system

FIGURE 6–13 _____
Oil burner in residential oil-fired furnace

Burners

Burners are the next major component to be discussed. A general description is given here to help readers understand the function of the burner in a unit, but details concerning specific burners are given in the chapters for each fuel.

The three main functions of any burner, regardless of the fuel used, are to

1. Meter the flow of fuel and air for combustion
2. Provide proper mixing of fuel and air for combustion
3. Supply the fuel-air mix to a burner face where the mix can be ignited

The first function, metering, is performed by a part called an orifice in a gas furnace. In an atmospheric burner orifices are located at the primary air opening end of the burner tubes, where the gas enters the tube. In a power gas burner metering is provided by an orifice in the gas line ahead of the burner face. In an oil burner, the metering device is called the nozzle. The nozzle is at the end of the oil line, at the burner face. A typical oil burner is shown in Figure 6-13. The burner face is inside the furnace and is not visible in the picture.

In a pulse-type burner, which is a burner used in some modern furnaces, an orifice in the gas line ahead of the burner face provides the metering function. This is similar to the metering orifice used in a power gas burner.

The second function, fuel and air mixing, occurs in the burner tube in an atmospheric burner used in gas furnaces and at the burner face in power gas burners, pulse burners, and in oil furnace burners.

The third function, supply of the fuel-air mix, takes place at the burner ports in a gas furnace and at the burner face in an oil furnace.

Although burners for different fuels vary in appearance and in the parts used to achieve the function, the result achieved is the same: safe and controlled combustion.

Heat Exchangers

The heat exchanger in a warm air furnace is the component that extracts heat from the flames and products of combustion and transfers that heat to the air that is used as the heating medium.

In a warm air system, it is necessary to separate the products of combustion from the air used to carry the heat. Products of combustion may contain harmful or dangerous gases that cannot be safely introduced into a habitable space. The heat exchanger is the part of the unit where this separation takes place. Figure 6-14 shows a heat exchanger that is used in a gas-fired, warm air furnace. The model shown has a curved design that increases the efficiency and ensures quiet operation. The furnace blower blows the air to be heated up between the sections of the exchanger where it is heated. The products of combustion go through each section and are completely separated from the air on the outside.

Gas furnace heat exchangers are normally made of cast iron or sheet steel. Most of them currently used are made of steel. The steel is formed into sections, and the sections are joined together to make the exchanger. Oil furnace heat exchangers are normally round, drumshaped sections. The burner flame fills the lower part, or fire pot, and the upper part serves as the exchanger. The air from the blower passes over the outside of fire pot and exchanger, and the flame and flue gas is contained inside.

Most high-efficiency furnaces utilize a secondary section of heat exchanger to extract heat from the products of combustion after they have passed through the regular heat exchanger. The secondary heat exchanger is located so the products of combustion leaving the regular heat exchanger pass through the secondary section before leaving the furnace. An illustration of both primary and secondary heat exchangers is shown in Figure 6-14a. The secondary section is also located downstream from the regular heat exchanger in the air stream pattern. The secondary heat exchanger removes additional heat from the products of combustion after they leave the regular heat exchanger. This heat increases the overall efficiency of the furnace. This also reduces the temperature of the flue gas leaving the unit. There is a great deal of water vapor in the products of combustion, and because of the lowered temperature of the products of combustion leaving the heat exchanger condensation can, and usually does, occur in the secondary heat exchanger. For this reason, the secondary heat exchanger is built of stainless steel, or other rustproof material, and has

FIGURE 6–14
Heat exchanger used in warm air furnace

SECONDARY HEAT EXCHANGER

FIGURE 6–14A ───────────────

A secondary heat exchanger is used in some furnaces to increase the heating capacity of the furnace

some means built in to it to drain the condensation. Also, most high efficiency furnaces use either power burners or draft inducing fans to facilitate flue gas removal.

Although there are many types and styles of heat exchangers, each one performs the same functions in the unit: they extract heat from the flames and products of combustion, and separate the products of combustion from the air used to heat the building.

6.3 SUMMARY

Proper control of the combustion process for the production of heat is necessary. This control is provided by burning fuel in a heating unit that produces the heat at a specified rate and also provides a medium to distribute the heat. Air and water are the most common mediums. A heating unit used to provide warm air for heating a building generally is called a furnace. In this chapter furnaces are covered.

A typical furnace is made up of a cabinet, blower, filter, heat exchanger, and necessary controls. The performance of each part is important in producing the desired results. The way each part interacts with the other parts makes the safe and controlled production of heat possible.

The cabinet of a typical furnace is used to hold the other parts together, and to direct the flow of air through the unit properly. The cabinet usually is manufactured of metal and then painted or enameled to provide protection for the unit, and to improve the looks.

The blower in a furnace normally is made up of a blower wheel, blower scroll, bearings and supports, and a motor to drive the wheel. Both belt-drive and direct-drive blowers are commonly used. A belt-drive blower has the blower motor mounted at the side of the blower and a belt and sheave arrangement provides a connection between them. A direct-drive blower has the motor mounted directly on the blower shaft and the blower wheel turns with the motor.

Filters in or adjacent to a furnace are used to filter the air that goes through the unit. Most filters furnished with furnaces are designed to provide filtering of large particles. Higher efficiency filters can be provided if desired.

The burner used in a given furnace is matched to the

type of fuel the furnace uses. The burner must produce the amount of heat needed by combining fuel and combustion air and burning the combination safely.

The heat exchanger used in a given furnace is designed for use with the burner. Two main types of heat exchangers are used, they are clam shell and drum types. Many variations of these two are also used, with some special types used in high-efficiency furnaces.

6.4 TERMS TO KNOW

cabinet
blower
centrifugal blower
axial-flow blower
impingement
filter efficiency
micron
mechnical filters
electronic filters
pressure drop
ionize
heat exchanger

CHAPTER
7
Gas-Fired Furnaces

7.1 INTRODUCTION

Natural gas is one of the most commonly used fuels, and natural gas fired furnaces are most commonly used. This chapter covers different types of natural gas furnaces. The chapter contains a general description of the parts and controls used in gas-fired furnaces, with typical applications for the different types of furnaces. Combustion safety controls also are covered.

Although natural gas is the most commonly used fuel gas, other gases are also used for heating. Most important among these are the liquified petroleum gases (LPG). Only minor modifications have to be made to a natural gas fired unit to change it to an LPG-fired unit. The descriptions of the furnaces in this chapter can apply to all gas-fired furnaces.

7.2 MAJOR PARTS

A gas-fired furnace has six major parts:

1. Cabinet or frame
2. Filter
3. Blower
4. Burners
5. Heat exchangers
6. Controls

The first three of these were covered in detail in Chapters 5 and 6; the controls will be covered in detail in a later chapter.

Burners

A furnace burner is the part of the unit where combustion takes place. The burner is designed to control combustion so heat is produced safely and at a fixed rate. Two basic types of burners are used in gas-fired heating equipment: atmospheric and power. Atmospheric burners generally are used in smaller units that are installed inside buildings. Power burners are used on some high-efficiency furnaces, larger units and especially those installed outside a building, such as in rooftop units.

Atmospheric Burners. Atmospheric burners, as the name implies, rely on atmospheric pressure to supply air for combustion. Figure 7-1 shows an atmospheric burner and indicates where air enters the burner tube at the front and also where the air is drawn into the flame. The air that mixes with the gas in the tube is called **primary air;** the air that is drawn in at the flame is **secondary air.**

The mixture of gas and air must be in the proper proportions for combustion to be achieved. The design of the burner is such that this mixture is achieved by natural draft, from the air surrounding the unit.

The description given here is of an individual burner. Some furnaces use one large atmospheric burner as a source of heat, but most use several burners fastened together into an assembly called a burner tray. When only one burner is used the heat exchanger usually is a drum type. In a furnace using several burners, the heat exchanger has a separate section for each burner. The number of burners and sections of the exchanger determines the heating capacity of the unit.

The construction of an atmospheric burner is simple and practical. See Figure 7-2 for an illustration of a burner. The burner is a tube, open at one end, where air and gas enter. The gas is brought to the open end of the burner tube through a pipe called a manifold and flows into the tube through a fitting called an **orifice.** A hole in the orifice is drilled to the proper size to meter the flow of the gas. The tube has a series of **burner ports** or openings along the top where the gas-air mix, now a combustible mixture, leaves the tube and is ignited.

When the gas enters the burner tube out of the orifice, it enters at a high enough velocity to create a low-pressure area around it. Air is **entrained** with the gas because it is pulled into this low-pressure area. The air that is entrained is the primary combustion air. Normally, there is a shutter on the open end of the tube, around the orifice. The shutter can be adjusted to regulate the amount of air drawn in so that the proper mix of gas and air is achieved. Figure 7-3 illustrates the air adjustment shutter located on the burner tube face.

The gas-air mix flows down the burner tube and out of the burner ports on the top of the tube. This mix is ignited by a pilot light or electric spark as it leaves the ports. Although the gas and air in the tube may be in the right

FIGURE 7-1 _____
Drawing of atmospheric burner showing how the gas and air mix and are carried to the burner ports where combustion occurs

FIGURE 7-2 _____
Parts of atmospheric burner

Primary air – Air that is mixed with fuel gas before firing.

Secondary air – Air that helps support combustion by mixing with the fuel gas in the actual burning process.

Orifice*

Burner ports(s)*

Entrained air – Air that is drawn into the stream of gas coming out of the orifice.

*Term defined in text.

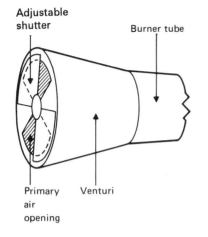

FIGURE 7-3 _____
Primary air adjustment shutter on atmospheric burner

FIGURE 7-4 _____
Typical gas valve

proportion to allow ignition, additional air for combustion is always necessary. This additional air flows in through the front of the burner vestibule and mixes with the gas at the flame during the actual combustion process. This air is the secondary air. A properly designed burner will achieve smooth, quiet, and efficient combustion.

The gas pressure for an atmospheric burner should be set at 3 1/2 in. wg for the burner to function properly. Also, the primary air has to be adjusted correctly, and there must be sufficient secondary air for complete combustion.

The flow of gas to the burner is controlled by a gas valve that is designed to control the flow of the gas to the unit burners. A typical gas valve used in conventional furnaces is shown in Figure 7-4. Many types of valves are used on heating equipment, but the basic operation of most of them is quite similar. Gas valves used with electronic controls, or with electric ignition, often have some special features that enhance their use on those systems.

A gas valve is located in the gas line ahead of the burner manifold. When a solenoid, which is part of the valve assembly, is energized, the valve opens and gas flows through it. When the solenoid is not energized, the valve is closed and no gas flows.

Some gas valves have built-in gas pressure regulators, and most have automatic pilot safety valves. If there is no pressure regulator in the valve, there will be one in the gas line ahead of it. There also will be a pilot safety valve in the line if it is not part of the valve assembly.

The function of the pressure regulator is to regulate the line gas pressure so that it is suitable for the orifices used in the burner. The pilot safety valve provides safety shutoff if the pilot is not lighted on a call for heat.

The gas pressure is regulated to the 3 1/2 in. wg for natural gas and 11 in. wg for LPG, at the pressure regulator by an adjusting screw under a protective cap on the stem of the regulator. The cap is removed and the screw turned to adjust the pressure. This action is done while the pressure in the gas line downstream from the regulator is checked with a manometer or a gauge.

After the pressure is set, the primary air is adjusted by turning the primary air shutter on the front of each burner. The adjustment is right when the flame burns evenly along the surface of the ports. The flame should be blue with just a touch of yellow in the tips. If there is too little primary air, the flame will be lazy. It will burn very softly and have a great deal of yellow in it. If the gas pressure is too high, or there is too much primary air the velocity of the gas-air mix through the ports can be great enough to cause the flame to burn up above the burner, causing a noisy flame.

If not enough secondary air is available at the burner location, the flame will appear very soft and lazy and will often burn out through the front of the burner ports. In some cases, the flames can roll out enough to burn the wiring and the controls in the **burner vestibule.** Usually, insufficient secondary air will occur when the furnace room is partitioned off and provisions are not made to bring air into it from other parts of the building or from outside.

Power Gas Burners. A power gas burner is one in which combustion air is provided by a blower instead of by atmospheric pressure. Gas is introduced through a firing head, at a given pressure. The firing head is normally shaped to direct the flow of the gas, and air from the combustion blower is introduced around the head. The diagram in Figure 7-5 illustrates the construction of a gas power burner.

A typical gas power burner has an orifice in the gas line leading to the burner head. This orifice is sized to control the flow rate of the gas and thus regulate the heat output of the unit. The burner face is a perforated, diverting nozzle. In some burners, it is round, and in others it may be square or rectangular. The gas flows out of the nozzle and mixes with the combustion air. This gas-air mix is ignited by an ignition device, such as a spark plug. Since the proper amount of air has to be mixed with the gas for good combustion, there is an adjustment on the combustion air intake, at the blower. The drawing in Figure 7-6 shows one type of air adjustment used on power burners.

In some heating units, an induced draft blower is used in the vent downstream from the combustion chamber outlet. This blower produces a draft in the vent which helps exhaust the products of combustion from the heating unit. This concept is used in some high-efficiency condensing-type furnaces.

Pulse Burner. A pulse burner, used in some high efficiency furnaces, is a low excess air burner. A fuel/air mixture is introduced at the burner face in short pulses or bursts. The pulses of the mixture are controlled by pressure in the firing chamber. The gas/air mixture is introduced to the firing chamber through a small flapper valve. The flapper valve opens to let gas into the burner, but when firing occurs back-pressure in the firing chamber closes the valve. The flapper remains closed until the pressure in the

Burner vestibule – The space in the front of the furnace where the burner and controls are mounted.

FIGURE 7-5 _____
Combustion air and gas mixing at the firing head in a gas power burner

FIGURE 7-6 _____
One type of combustion air adjustment used on a power burner

FIGURE 7-6a _____
A special burner is used in a pulse-type furnace to increase the combustion efficiency of the furnace

Light-sensitive cell – Any one of several types of photoelectric cells used for flame sensing.

Flame rod – An electric conductor used as a flame-sensing device.

firing chamber drops as the products of combustion leave the chamber. The burning cycles of a pulse burner are short but continous. A pulse burner is shown in Figure 7-6a.

A pulse burner operates more efficiently than conventional furnaces because only the amount of combustion air necessary for complete combustion is used by the burner.

The complete power burner assembly has a gas valve, a pressure regulator, and ignition safety controls. The valve and the pressure regulator are similar to those used with an atmospheric burner, but a standing pilot usually is not used with a power burner, so a pilot safety is not needed. In this type of safety, a **light-sensitive cell** or **flame-rod** type of sensor monitors the fire, both on start-up and continually during a call for heat.

The light-sensitive type of safety is similar to that used on an oil-fired burner. This safety is described in detail in Chapter 8. The flame-rod safety has an electric conductor located so that it is in the flame when the burner is operating. When the fire is burning, the flame and the conductor form an electric circuit. The flame will conduct electricity because of the carbon particles in the fuel and the unburned particles of carbon in the flame. This circuit is part of a safety circuit, and the flow of electric current through it is monitored by an electronic control device. If on a call for heat, the flame does not come on within a time considered safe for combustion, the control device will shut the unit down on safety. The control device will also shut the unit off "on safety" if the fire goes out during a firing cycle.

A power burner can be designed as a two-stage burner, with a low-fire and a high-fire stage. A two-stage burner has one-half the heat output at low stage as it does at high stage. This difference allows a designer to select a unit that will more nearly match the heat load on a building at two different load conditions. When a two-stage burner is selected, two gas valves are used; one controls gas flow for low-stage fire, and both are used for high fire. When a unit is two-staged, the combustion air flow rate has to be regulated for each stage. There must be less air for low stage and more for high. This regulation is accomplished by using a motorized damper on the combustion air intake. When the thermostat calls for one-stage fire, the damper is opened the proper amount, and on a call for two stage, the damper opens more.

Power burners usually are used in larger furnaces or in units that are located outside a building. When a heating unit is installed outside a building, wind may affect the air pressure around the unit. In such a case, an atmospheric burner is hard to control. Because a power burner is not dependent upon atmospheric pressure for its operation, a power burner can be used outside as well as inside.

Heat Exchangers

The heat exchanger is one of the most important components of the heating unit. Its purpose is to allow the exchange of heat from the products of combustion to the air.

The design of the exchanger varies somewhat depending on the burner used. Two basic types of heat exchangers are used in conventional warm air heating equipment: the clamshell and the drum type. In some furnaces secondary sections are added to increase the efficiency of the heat exchanger, or specifically for use in high-efficiency furnaces. The clamshell exchanger normally is used with atmospheric burners, and the drum type is used with power burners.

Materials Used in Heat Exchangers.   Most of the heat exchangers manufactured today are made of sheet steel. Cast iron was used in the past, but because of the greater mass of material in cast iron, heating of the sections is slower on warm up, and, consequently, cooling of the metal also is slower on an off cycle. This slowness is an advantage in furnaces without blowers because of the desirability to have a constant, steady source of heat for heating the air. In today's forced-air systems rapid transfer of heat is more desirable, so lighter and thinner metal is used in the exchangers.

Special steels or special coatings often are used in the construction of heat exchangers. The special materials make the exchanger more resistant to corrosion caused by acids produced by the products of combustion, or to burn-outs from the flame.

Aluminized steel often is used for the construction of exchangers. Aluminized steel is steel treated in a process that bonds a thin coat of aluminum to the outer surface of the steel. This process makes the steel more resistant to corrosion.

Glass or ceramic-coated steel also is used for exchangers. The coating in this case is a thin layer of glass or ceramic material that is melted onto the steel exchangers. The glass or ceramic makes an impervious coating on the steel that will resist corrosion as long as the coating is intact.

Stainless steel is another material used for heat exchangers when extra durability is desired. Stainless steel is resistant to most of the types of corrosion that may occur with heat exchangers. It is an excellent material for heat exchanger construction.

All of the special coatings or materials used for heat exchangers are more expensive than plain steel and add to the cost of the original unit. In applications where corrosion may be a special problem, however, the cost difference may easily be offset by replacement cost. In some cases, manufacturers give a better warranty on the exchanger made out of special materials than they do on the plain steel one.

Clamshell Exchanger.   The clamshell exchanger is the most common type of heat exchanger. It is used in smaller

Clamshell heat exchanger

Clamshell
heat
exchanger

FIGURE 7-7
Clamshell heat exchanger and cutaway of downflow furnace with exchanger installed

gas-fired furnaces almost exclusively. Drawings of a typical heat exchanger and a downflow furnace with the heat exchanger installed are shown in Figure 7-7.

The clamshell exchanger consists of multiple sections, or "clams." Each is formed of two pieces of steel welded together in such a way that they form a tight chamber inside. The burners are then inserted in these sections, at the bottom, as shown in Figure 7-8.

A number of sections are joined together to make up an exchanger. In most furnaces, each section is designed for an input of approximately 25,000 Btu/h of heat. The capacity of a furnace can be estimated by multiplying the number of sections in the exchanger by 25,000. The result is the approximate Btu/h input capacity of the furnace.

When the burners are fired, the flames and products of combustion rise through the inside of the clamshell sections, and the flue gases are exhausted at the top of each section through an opening that leads to the flue. The air that is being heated passes up through the openings between the clamshell sections. Figure 7-9 shows a diagram with the airflow around the outside of the sections of a heat exchanger.

As the burner cycles on and off, the various sections of the heat exchanger heat and cool. In this process the metal expands and contracts, so provisions must be made in the construction of the clamshells to keep the metal from distorting. The usual arrangement consists of internal ties and braces or **spot welds** to prevent this distortion. These ties make the clamshell somewhat rigid. If the metal is restrained too rigidly, it will bend at the points of attachment and will break due to the continuous movement. Also, if the attachments are too rigid, the heat exchanger is noisy because of stresses created at points of connection. Popping or cracking sounds are created when the exchanger is heating up or cooling down. Manufacturers have to provide for some movement of the heat exchanger in the unit to prevent this noise, or breaking of the metal due to **metal fatigue.**

Drum-Type Exchanger. A second type of heat exchanger used is the drum type. The drum type of heat exchanger is normally used with a power burner. The burner is housed in the lower part of the heat exchanger, which is called the **combustion chamber.** The combustion chamber is made

FIGURE 7-8 ⸺
Clamshell heat exchanger with cutaway showing burner installed

Spot weld – A method of welding two pieces of thin metal together by clamping them between two metal electrodes and passing an electric current through the metals between the electrodes.

Metal fatigue – Breaking of metal caused by continued bending and flexing.

FIGURE 7-9 _____
Airflow in clamshell heat exchanger

FIGURE 7-10 _____
Drum-type heat exchanger

of heavy gauge metal and is lined with **refractory** materials. The combustion chamber actually is part of the exchanger. Figure 7-10 shows a typical design for a drum-type exchanger. The flames from the burner are contained in the combustion chamber, and the hot products of combustion rise out of the pot into a larger section at the top. The hot products then flow through a connection to the flue pipe. The air to be heated flows around the outside of the combustion chamber and the upper section, where it picks up heat from the hot surfaces. The hot air then is used for heating the building.

Some drum-type heat exchangers have an additional section where the products of combustion flow through secondary tubes after passing though the main exchanger. In such a heat exchanger more hot metal surface is exposed to the air. This type of exchanger is called a primary-secondary, or multipass, exchanger. A sectional drawing showing the construction of a multipass exchanger appears in Figure 7-11.

Most high-efficiency furnaces use a secondary heat exchanger to increase the output capacity of the furnace. The heat exchanger may be of the same general type as the

Refractory – A material that will not melt at the operating temperature of the burner; a device used to separate the heat exchanger surface from the flames in a furnace.

FIGURE 7-11 _____
Cutaway view of multipass heat exchanger

FIGURE 7-11a _____
Most high efficiency furnaces use a secondary heat exchanger to increase the efficiency of the unit

primary exchanger, or it may be entirely different.

In some clamshell-type heat exchangers extra clamshells are used as the secondary section of the exchanger. Figure 7-11a shows a furnace with a secondary heat exchanger. In other furnaces the secondary section is more like a coil used in air conditioners, only with larger tubes in the coil. In any case, the secondary sections must be made of a material that will not be corroded or damaged by water or water vapor in the flue gas. As extra heat is extracted from the flue gases in a high-efficiency furnace, the flue gas temperature drops below the dew point temperature. This causes water vapor in the flue gas to condense in the exchanger. Means must be made to drain the water from the heat exchanger. Since the flue gas leaving the secondary heat exchanger usually is not hot enough to produce draft in the chimney carrying the flue gas away, a power burner or induced draft fan is used on the unit. One advantage to this is that a flue pipe that is much smaller than that used on a conventional furnace can be used.

Flue Connections.   The basic function of the heat exchanger is to keep the products of combustion separate from the air being heated. Thus, the heat exchanger must be connected to a vent or a flue to exhaust the flue gases.

FIGURE 7-12
Flue gas leaving heat exchanger on a gas furnace

Dilution air – Air that is allowed to mix with the products of combustion after they leave the heat exchanger.

Openings are made in the top of the clamshells or drum and then connected to the flue. In an atmospheric burner, the flue openings from the clamshells are connected to a device called a draft diverter. The draft diverter connects to the flue. On a drum-type exchanger, the connection is made directly to the flue; however, a device called a barometric damper is used in the flue at the connection to the furnace, that maintains a proper draft across the fire in the combustion chamber. Figure 7-12 shows the path that the flue gas takes as it leaves a clamshell exchanger and goes into the flue or vent.

A draft diverter, or a barometric damper used with a power burner, is used to allow **dilution air** to mix with the products of combustion and to prevent the natural draft effect from pulling the flames out of a furnace combustion chamber.

Proper care should be taken in the sizing of vents or flues external to a furnace. It is important to maintain a high enough temperature in the sections of vent to prevent excessive condensation, and to maintain a high enough velocity of flue gases to create a draft. Many of the newer high-efficiency furnaces use very small vents, but the external sections still must be properly designed. Furnace manufacturers and vent manufacturers publish sizing data which should be used.

Heating Capacity

Gas-fired furnaces are available in many different sizes and heating capacities. Manufacturers make different sizes of furnaces available so designers can better match the heating load loss for any given job. To provide different heat capacities, the manufacturer has to consider three factors:

1. Heat value of the gas to be used
2. Flow rate of the gas into the unit
3. Efficiency of the heat exchanger

Each of these factors is covered in the following sections.

Heat Value of Gas.   The heat value of the gas is determined by the type of gas to be fired. Natural gas, the most common of the gases used, has a nominal value of 1000 Btu per cubic foot. For every cubic foot of gas that the furnace uses, 1000 Btu of heat will be produced. A 100,000 Btu/h furnace burns 100 cu ft of gas per hour.

Liquid petroleum gases (LPG) come in the liquid state and are rated in either Btu/gal or Btu/cu ft. The heat value of these gases per gallon varies as the chemical composition varies. The nominal heat value for butane is 3200 Btu/cu ft of fuel, and for propane it is 2500 Btu/cu ft. Values of other gases are shown in tables found in Figure 2-1 in Chapter 2.

Gas Flow Rate. The flow rate of the gas into the burner is controlled by the pressure of the gas in the line and by the size of the hole drilled in the orifice at each burner head. The gas pressure is controlled by a pressure regulator. Units are commonly designed to produce a specific amount of heat per burner at a given pressure. This means that each orifice is sized to allow the quantity of gas to flow per hour that will produce that amount of heat exactly. The hole in the orifice is sized according to the size shown on an orifice sizing chart. One is found in Table A-2 in the Appendix. This chart shows the orifice size required for specific flow rates at various gas pressures.

Efficiency. The third factor involved in heating capacity is the efficiency of the burner and the heat exchanger combination used in the furnace. The efficiency rating of a unit gives the amount of heat available for heating a building as a percentage of that produced in the combustion process. The normal efficiency factor is 80 percent for a conventional furnace, but ranges much higher for high-efficiency furnaces. In conventional furnaces some heat has to be allowed to go up the flue to provide draft. About 20 percent of the total produced goes up the flue, so 80 percent is left for heating. The amount of heat produced in the combustion process is called the **input** of the furnace, and the amount left over after deration is called the **output**.

Unit Sizes. Gas furnaces are manufactured in many different physical sizes and heating capacities in each of the styles available. The physical sizes are determined by the capacity of each unit. The capacities for residential furnaces start at about 50,000 Btu/h. The capacities increase in increments of approximately 25,000 Btu/h, up to 300,000 Btu/h. Much larger units are made for commercial or industrial applications. Figure 7-13 is a page from a manufacturer's catalog showing some of the sizes available in one type of furnace.

Input – Heat, in Btu/h, produced by a furnace as related to fuel use.

Output – Actual heat, in Btu/h, available for heating after a furnace has been derated for efficiency loss.

PP/PRODUCT PRESENTATION ——————————————— HEATING

RESIDENTIAL TYPE SINGLE ZONE HEATING

Up-flo, down-flo and horizontal models are available in gas, oil or electric heat. Most are designed to accept air conditioning. A large selection of evaporator coils, outdoor condensing units, electronic air cleaners and humidifiers can be added for economical all-season total comfort.

All models are equipped with either resiliently mounted belt drive or multispeed beltless blowers. A durable, enamel finish is baked-on for lasting good looks. Matching return air cabinets have been designed to compliment up-flo installations.

Gas furnaces feature the exclusive Lennox DURACUR-VE ® heat exchanger. It's revolutionary design reduces metal fatigue, ticking and resonance. An optional DURAGLASS II ® coating helps fight corrosion for added heat exchanger life. Quiet continuous-slot burners provide complete combustion. They end irritating popping and flashback.

Electric furnaces contain strong nichrome heating elements with higher temperature capabilities and longer life. They are placed directly in the airstream to give instantaneous heat transfer — no waste. Completely automatic controls efficiently sequence elements. Double protection is provided by circuit breakers and individual element temperature limit controls.

Conservator Electric ignition and optional vent damper for highest gas heat efficiency. Also on G12 "E" models.

GAS UP-FLO MODELS

Model Number *	Input Btu/h	Tons of Add-on Cooling (in.)	Flue Size (in.)	Net Weight (lbs.)	Dimensions (in.)		
					H	W	D
G12D2-40	40,000	1½ or 2	4	143	49	16¼	26⅛
G12D2E-40	40,000	1½ or 2	4	143	49	16¼	26⅛
G12D2-55	55,000	1½ or 2	4	143	49	16¼	26⅛
G12D2E-55	55,000	1½ or 2	4	143	49	16¼	26⅛
G12D2-82	82,000	1½ or 2	5	158	49	16¼	26⅛
G12D2E-82	82,000	1½ or 2	5	158	49	16¼	26⅛
▲ G11Q3E-82V Conservator	82,000	2, 2½ or 3	4	167	49	16¼	26⅛
G12Q3-82	82,000	2, 2½ or 3	4	166	49	16¼	26⅛
. G81-85-110	85,000 110,000	2½ thru 5	5	275	53	26	26⅛
G12D-110	110,000	----	5	185	49	21¼	26⅛
▲ G11Q3E-110	110,000	2 thru 3	5	194	49	21¼	26⅛
G12Q3-110	110,000	2 thru 3	5	192	49	21¼	26⅛
G12Q3E-110	110,000	2 thru 3	5	192	49	21¼	26⅛
G12Q4-110	110,000	3½ or 4	5	200	49	21¼	26⅛
G12Q4E-110	110,000	3½ or 4	5	200	49	21¼	26⅛
▲ G11Q3E-110V Conservator	110,000	2½ and 3	5 oval	194	49	21¼	26⅛
▲ G11E-110V Conservator	110,000	2, 2½ or 3	5 oval	193	49	21¼	26⅛
G12Q3-137	137,000	2½ and 3	6	236	53	26¼	26⅛
G12Q3E-137	137,000	2½ and 3	6	236	53	26¼	26⅛
G11Q4E-137	137,000	3½ or 4	6	252	53	26¼	26⅛
G12Q4-137	137,000	3½ or 4	6	252	53	26¼	26⅛
G12Q4E-137	137,000	3½ or 4	6	252	53	26¼	26⅛
G12Q5-137	137,000	3½ thru 5	6	262	53	26¼	26⅛
G12Q5E-137	137,000	3½ thru 5	6	262	53	26¼	26⅛
G11Q3E-137V Conservator	137,000	2½ or 3	6 oval	239	53	26¼	26⅛
▲ G11E-137V Conservator	137,000	3, 3½, 4 or 5	6 oval	249	53	26¼	26⅛
G12Q5-165	165,000	3½ thru 5	6	278	53	31¼	26⅛
G12Q5E-165	165,000	3½ thru 5	6	278	53	31¼	26⅛
▲ G11Q5E-165V Conservator	165,000	4 or 5	6 oval	301	53	31¼	26⅛
▲ G11E-165V Conservator	165,000	3, 3½, 4 or 5	6 oval	278	53	31¼	26⅛
G81-165	165,000	6	6	520	58	38	28
G11E-200V	200,000	3½, 4 or 5	7	291	53	31¼	26⅛
G81-220	220,000	7½ or 11	(2) 5	675	58	51½	28

▲ Also available in 50 Hz models.

FIGURE 7-13 ———————————————————————————————

Page from manufacturer's catalog showing some of the sizes of units available

Controls

Controls on a gas-fired furnace are used to operate the furnace in such a way that the amount of heat is produced to heat a building or building spaces to a desired temperature to match the heat loss in those spaces. The controls used on a gas-fired furnace include a thermostat, limit controls, blower control, and combustion safety controls.

Thermostat.   A gas-fired furnace thermostat is basically a temperature-actuated electrical switch. It contains a bi-metal sensing element that opens or closes a set of electrical contacts. The sensing element and switch are mounted on a movable plate that is attached to an adjustment lever. The adjustment lever permits selection of a setpoint at which the switch opens and closes. The electrical switch is wired into the control circuit on a furnace.

Limit Control.   A typical limit control on a gas-fired furnace is a low voltage limit located in the control circuit between the transformer and the gas valve on the unit. The limit control is a temperature-actuated electrical switch. A bi-metal sensing element opens a set of electrical contacts if the temperature goes above a certain setpoint. The sensor for the limit control is inserted in the furnace where it senses air temperature either by the heat exchanger or in the warm air plenum. If the limit control opens on a temperature rise above its setpoint, the control circuit is disconnected and the unit shuts down.

Blower Control.   A blower control is used on a typical gas-fired furnace to control the blower motor. A typical blower control has a bi-metal element that opens and closes a set of electrical contacts on change of temperature. The control is adjusted to close the contacts when the temperature surrounding the bi-metal reaches a predetermined setpoint between normal room temperature and about 140° F. The bi-metal is inserted in a furnace into the air passing over the heat exchanger or in the hot air plenum.

Some heating unit control systems include time delay relays to bring on the blower after a given time delay. These controls are often used on outdoor package heating units.

Thermocouple – A device that when heated produces a low-voltage electric signal.

Thermistor – A solid-state device in which the resistance to an electric current varies as the temperature varies.

FIGURE 7-14 _____
Types of thermocouples available from one manufacturer for pilot safety application

Combustion Safety Controls

To ensure that the fuel-air mix on a gas-fired furnace is fired as soon as it is introduced to a burner, combustion safety controls are used. Because the fuel/air mix is explosive, it must be fired as soon as it is introduced. Any delay in firing could cause a buildup of the mixture, and if an accumulation was fired, the result could be a disastrous explosion. The furnace is designed to burn fuel at a steady, sustained rate.

Combustion safety controls monitor the combustion process to make sure that firing takes place within a reasonable time after gas starts to flow. If firing does not take place, the control shuts the gas valve; or if the flame goes out during a firing period, the valve also is shut off.

The two most common types of combustion safety controls used on furnaces are (1) pilot safety and (2) flame surveillance. Some pilot safety controls use a standing pilot while others use a pilot light that is only lighted on a call for heating.

Pilot Safety Controls. Before the advent of electronic control units, most gas-fired furnaces using atmospheric burners utilized a standing pilot, or a pilot flame that was constantly burning, to fire the gas-air mix on a call for heat. In such systems, it is necessary to provide some means of monitoring the pilot flame to make sure it is burning, so the gas will be ignited when the gas valve opens. The device that is used most commonly to do this monitoring on older style, conventional furnaces is called a **thermocouple.** The thermocouple is used with a safety valve. The thermocouple is located so the pilot flame heats the end of the device when the flame is lighted and controls the safety valve. Figure 7-14 shows some of the types of thermocouples available from one manufacturer.

When a thermocouple is heated, it generates a low-voltage electrical signal. This signal is conducted to a magnetic coil in a safety shutoff valve located in the gas line ahead of the control valve. In some cases, this shutoff valve is part of the main gas valve itself. A typical gas valve with the safety shutoff coil and valve is shown in Figure 7-15.

When the electric coil in a shutoff device is energized by the electric current from a thermocouple, gas can flow through the valve. If the pilot flame is out, no current is generated, and the shutoff valve will not open. As a result gas will not flow into the burners.

There are other less commonly used types of sensing devices for flame monitoring. Among them are warp switches and remote bulb temperature-pressure devices. In a warp switch, heat from the flame is converted directly into mechanical action, which is used to control a valve. In the remote bulb type of control, the heat causes pressure in a tube, and the pressure is used to generate a mechanical motion, which controls a valve.

Electric Ignition Safety.   Electric ignition safety devices are used on most gas-fired furnaces now being manufactured. They are commonly used with electronically controlled units. This type of ignition system ignites the gas-air mix with an electrical spark. This ignition system is often used to light a pilot light on call for heat, and the pilot light then burns only until the heating call is satisfied. In larger units the electric spark may be used to fire a gas/air mixture directly. The ignition system may be part of the original ignition system on a furnace, or it may be used as a conversion on a conventional gas furnace.

When electric ignition is used, a flame-proving method of ignition safety is used. This method is usually one of three types: (1) a bimetal warp switch, (2) a thermocouple generating a low-voltage electric signal, or (3) a solid-state system using flame rectification for sensing.

Power Burner Combustion Safety.   When a power burner is used, the flame must be monitored both on start-up and during the "on" cycle. On a call for heat, a combustible fuel-air mixture enters the combustion chamber. If ignition does not take place, the accumulation of this mix is highly dangerous. Ignition of a large quantity of fuel and air would cause an explosion.

The most common method of monitoring ignition when a power burner is used is by a type of cell that "sees" the flame when the burner is on and that controls an electric signal generated by a control center. The electric signal, in turn, is used to control the main valve or operating relay for the burner.

A cell that is used to scan the flame is an electronic device in which the resistance to electrical flow varies with the amount of light the cell absorbs. Figure 7-16 is a photograph of a light cell showing the mounting and connecting

FIGURE 7-15 _____
Gas valve with pilot safety built in

FIGURE 7-16 _____
Light cell, mounting bracket, and connecting wires used on a combustion safety device

Cover

Manual reset button

Terminal connectors for leads from cell

Mounting base

FIGURE 7-17
Control center with light-sensing type of sensor

wires. The more light that strikes the cell, the lower the resistance to the flow of electricity; the less light that strikes the cell, the higher the resistance. It is possible to monitor a flame by sending a low-voltage electrical signal through the cell and metering the amount of current flowing. An electronic control center generates the low-voltage signal, monitors the current, and controls the electric signal to the gas valve. So the operation of the gas valve is dependent upon the signal received as a result of light striking the sensor. Figure 7-17 shows a control center used with a light sensing type of sensor.

7.3 SUMMARY

All gas-fired furnaces, regardless of the particular gas fuel used, have many similarities. The fuel must be mixed with the air in the right proportions to burn efficiently, and provisions must be made to control the combustion process. Also, the furnace has to be built so that heat from the combustion process can be extracted from the flames and products of combustion and used to heat air.

The two main components of the furnace that produce the heat, and extract it, are the burners and the heat exchanger.

Two types of burners are used. The first is the atmospheric burner. This burner gets all of the combustion air from the atmosphere by natural draft. The other type of burner is the power type. This burner uses a blower to provide the combustion air. Each type of burner has special characteristics that make it more practical than the other to use in some circumstances.

Also, two main types of heat exchangers are used for gas-fired furnaces: the clamshell and drum-type exchangers. The clamshell exchanger normally is used with the atmospheric burner and has separate sections for each burner to fit into. The drum-type exchanger has a combustion chamber where the flame is produced by the burner and a separate section where most of the heat is transferred to the air. This type of exchanger is most often used with a power burner.

Some type of ignition safety control must be provided on each furnace. When the air-gas mix is introduced into a furnace through the burner, the mixture is explosive. The mixture must be fired at a continuous rate, or it can create a highly dangerous condition. Since various means are used to fire the gas-air mix, there are different types of flame proving, or combustion safety, devices used to ensure the ignition takes place. Among these are warp switches, thermocouples, and flame surveillance. Each device is used on various types of heating units. The choice of which to use depends upon the type of burner used and the application for which the furnace is designed.

7.4 TERMS TO KNOW

primary air
secondary air
orifice
burner
burner ports
entrained air
atmospheric burners
power gas burners
pulse burner
burner vestibule
light-sensitive cell
flame rod
clamshell exchanger
spot weld
metal fatigue
fire pot
refractory
combustion chamber
dilution air
input
output
thermostat
limit control
blower control
thermocouple
thermistor

CHAPTER

8

Oil-Fired Furnaces

8.1 INTRODUCTION

Oil-burning heating equipment is common in residential, industrial, and commercial applications. Although many of the components used in the manufacture of oil furnaces are similar to those used in gas furnaces, there is a difference in the oil burners and the heat exchangers.

Since the burner in an oil furnace is of a special type, the heat exchanger is special too. In this chapter the heat exchanger used most often in oil-burning equipment is described, and the function of the exchanger in relation to the burner is also described. The chapter also covers unit controls including types of combustion safety controls used to ensure proper firing of the oil-air mix in the oil burner.

8.2 MAJOR PARTS

Oil-fired heating units are manufactured in many different styles and varieties (such as upflow, downflow, and horizontal). Figure 8-1 shows these major configurations.

Basically, an oil burner is a power burner. That is, it does

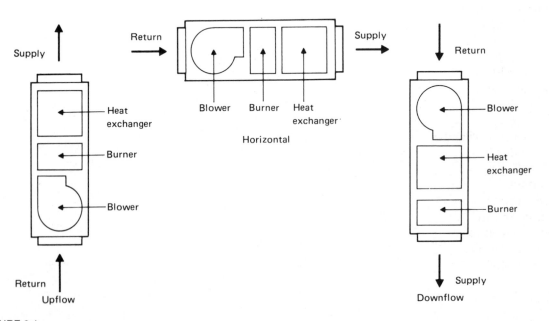

FIGURE 8-1
Three normal configurations of furnaces

not get its combustion air strictly by atmospheric pressure; rather, the air is supplied by a combustion air blower. Since the oil is in a liquid state, but can be burned only in the vapor state, one of the main functions of the burner is to vaporize the oil and then mix it with air for burning.

Oil Burners

The majority of burners used in oil furnaces are power burners. Some small units, and especially room heaters, use **pot-type burners,** but these are seldom used in furnaces and will not be covered in this text. Some low-pressure burners have been used and are still in use by various furnace manufacturers. A photograph of a typical oil burner used in an oil-fired furnace is shown in Figure 8-2.

The oil burner assembly is made up of various parts. These parts are: an oil pump, combustion air blower, gun assembly, blast tube, burner face, and ignition assembly. Each of these parts, in turn, is made up of several more parts.

Oil is in a liquid state when it is introduced at the burner head. Since oil will not mix with air for combustion in this state, the oil must be broken down into small droplets at the burner head and mixed with air. The breaking-down process is called **atomization.**

Atomization is one of the functions of the burner. Oil is delivered to the **burner nozzle** at a fairly high pressure from a pump that also is part of the burner assembly. Air is delivered to the firing head from a combustion air blower. As the oil passes through the burner nozzle, it is broken up into very small droplets. Air from the combustion air blower goes through the blast tube of the burner, and goes out of perforations in the firing head around the burner nozzle. The air mixes with the oil spray coming out of the nozzle. The air/oil mix is sprayed in a cone-shaped pattern into the furnace's combustion chamber. Figure 8-3 shows how the oil is atomized and mixed with air at the burner face.

The **oil pump** is a rotary pump that brings oil from the storage tank and delivers it to the nozzle at the burner head at a pressure of 100 lb per sq in. (psi). The oil pump is motor driven, normally by the same motor used to drive the combustion blower. The pump is self-lubricating and adjustable for outlet pressure. There is an adjustment screw on the pump and a port for a pressure gauge.

FIGURE 8-2

Typical oil burner suitable for a residential furnace

Pot-type burner – A burner for liquid fuels in which the fuel flows into a vessel where it is vaporized by heat and burned in the same vessel.

Atomization*

Burner nozzle – A steel nozzle through which fuel oil is introduced at the burner face. (The oil is partially vaporized, and the rate of flow is metered.)

Oil pump*

*Term defined in text.

FIGURE 8-3 _____
Atomized oil and air mixing at the firing head of
an oil burner for proper combustion

Combustion assembly – The oil line,
burner nozzle, and ignition electrodes.
(These parts are mounted inside the blast
tube and connect the oil pump to the
burner face.)

Bast tube*

Burner face*

Electrodes – Steel rods that conduct electricity from an ignition transformer and
produce a spark for ignition of the oil-air
mix.

Ignition assembly*

A **combustion air blower** is a small centrifugal blower
that provides air to the blast tube assembly so the air is
available at the burner face. The combustion blower may
be driven by the same motor that drives the oil pump.
Because the proportion of air to oil vapor is critical, the
combustion air blower can be adjusted to regulate the
amount of air the blower puts out.

The **gun assembly** includes the oil pipe that runs from
the oil pump to the nozzle at the burner face. The pipe
goes through the center of the blast tube. The oil burner
nozzle fits on the end of this pipe. The support bracket for
the gun assembly is adjustable so that the nozzle can be
positioned correctly in the burner face.

The large tube that carries the air from the combustion
blower to the burner face is called the **blast tube.** The gun
assembly fits inside it. The tube goes through an opening
in the combustion chamber wall and is the connection
between the oil burner outside the unit and the burner face
inside.

The perforated plate on the inside end of the blast tube
is the **burner face.** Combustion air passes through perforations in the plate, and the burner nozzle is located in a hole
in the center. Ignition **electrodes** are located just above and
in front of the nozzle. The oil-air mix is ignited as it enters
the combustion chamber.

The **ignition assembly** includes an ignition transformer that raises the voltage of the electricity coming to
the unit for supplying a spark for ignition of the oil-air mix
in the combustion chamber. Electrical conductors run from
the transformer to the ignition electrodes. On a call for
heat the oil pump, combustion air blower and the ignition
system are all electrically energized at the same time. As
oil leaves the nozzle and is mixed with the air from the
burner face, an electrical spark is generated between the
two ignition electrodes, and the oil-air mix is ignited. The
spark occurs adjacent to the cone of oil-air but not in it, as
shown in Figure 8-4.

The flame that is produced by the oil burner is a con-
centrated flame; it appears in one general locality.
Consequently, the flame can be contained in a combustion
chamber where the hot lining material helps produce the
high temperature needed for efficient combustion.

High-Efficiency Oil Burner.   A high-efficiency oil burner
normally is a burner with physical features that either

cause a more efficient mixing of the air and atomized oil, or recycles unburned oil and vapor mix through the flame for better combustion.

The more efficiently combustion air and atomized oil is mixed before it is burned, the more efficient the fire will be. If the fuel oil is not atomized, or broken up into fine particles in a burner head, it will not mix thoroughly with combustion air. Larger particles of fuel oil will not be burned completely in the flame, and a yellow flame, flame with many particles of unburned carbon, results. The more thoroughly the oil can be atomized the more like a gaseous fuel the air and oil mixture will be when it reaches the combustion chamber. To achieve more complete atomization special nozzles are used in some cases, in others, the burner head itself is designed to provide a better mix of air and oil. Higher pressure burners often are used to provide more efficient atomization.

Some oil burner firing heads are designed to recycle unburned fuel oil/air mixtures that enter the combustion chamber around the outside of the oil/air mixture cone in a combustion chamber. This mix usually has oil drops in it that have not been completely atomized or mixed with the air properly. A firing head is used that sets up a swirling air oil mix around the side of the flame in the combustion chamber that carries unburned oil and air back into the areas of the flame where the mixing originally occurred. Figure 8-4a shows the recycling pattern achieved in a "blue flame" burner.

High-efficiency oil burners usually provide a flame with less yellow in it than a regular high pressure burner. For this reason they are often called "blue flame" burners.

Heat Exchangers

Oil-fired heat exchangers used with power burners are wraparound, or drum type, as shown in Figure 8-5. Since a power burner produces a single flame, the heat exchanger used with it is designed to perform best with this flame pattern.

In an oil-fired heat exchanger, the combustion process takes place in a combustion chamber. The combustion chamber is the lower part of the exchanger and is lined with ceramic or fire-clay material. This material reduces the possibility of the flame's burning out that part of the

FIGURE 8-4
Ignition of the oil-air mix in an oil burner

Blue flame oil burner

Recirculated air & oil vapor

Blast tube

Flame

Oil burner

Recirculated air & oil vapor

FIGURE 8-4a _____
Some oil burners are designed to recycle unburned oil/air vapors to
increase the combustion efficiency

FIGURE 8-5 _____
Wrap-around or drum-type heat exchanger

heat exchanger and also produces an area of intense heat,
which helps the combustion process.

After firing, the hot products of combustion rise out of
the combustion chamber area and go into the heat
exchanger proper. This part of the exchanger is called the
drum. Here the metal is heated by the hot products of
combustion from the inside, while the air blown over the
outside surface is heated. This hot air is used to heat the
building. After passing through the drum, the products of
combustion go into a breeching, or vent, connection, which
is connected to the flue or chimney.

Some drum-type heat exchangers have a secondary
section where the hot flue gases pass through additional
tubes after leaving the drum, or primary section. See
Figure 8-6. A heat exchanger with a secondary section is
called a two-pass exchanger. The main advantage of this
type of exchanger is that more heat can be extracted from
the flue gases with a smaller exchanger.

Drum*

Draft diverter

Secondary
heating
surface

Refractory

Primary
heating
surface

FIGURE 8-6

Cutaway drawing of oil heat exchanger showing the secondary heating section in relation to
the primary section

Heating Capacity

The heating capacity of an oil furnace (rated in Btu/h) is
the function of the heat value of the oil burned, the rate of
firing the oil, and the efficiency of the unit. Combustion
and heat exchanger efficiency are figured together and are
determined by combustion efficiency tests that include the
performance of both.

Unit Operation.   Oil, being a liquid fuel, is measured in
gallons, and the flow rate of oil into a burner is rated in
gallons per hour (gph). The oil is pumped to the burner at
a pressure of 100 psi and introduced at the firing head
through a nozzle that controls the flow rate and breaks the
oil down into small droplets to facilitate its mixing with
air. The nozzle is sized to allow the oil to flow at a given
rate—usually from about 0.65 gph to 1.50 gph for residential and smaller commercial units. Larger nozzles are used
for large commercial and industrial units.

TABLE 8–1

Description and Specification of Common Grades of Heating Oil

Grade of Fuel Oil[b]	Flash Point F	Pour Point F	Water and Sediment % by Vol.	Carbon Residue on 10% Residuum %	Ash % by Wt.	Distillation Temperatures F			Saybolt Viscosity Seconds				Kinematic Viscosity Centistokes			
						10% Point	90% Point	End Point	Universal at 100 F		Furol at 122 F		at 100 F		at 122 F	
	Min.	Max.	Max.	Max.	Max.	Max.	Max.	Max.	Max.	Min.	Max.	Min.	Max.	Min.	Max.	Min.
1. A distillate oil intended for vaporizing pot-type burners and other burners requiring this grade[d] API gravity 35 (min.)	100 or Legal	0	Trace	0.15		420		625					2.2	1.4		
2. A distillate oil for general purpose domestic heating for use in burners not requiring No. 1 API gravity 26 (min.)	100 or Legal	20[e]	0.10	0.35		[e]	675		40				(4.3)			
4. An oil for burner installations not equipped with preheating facilities	130 or Legal	20	0.50		0.10				125	45			(26.4)	5.8		
5. A residual type oil for burner installations equipped with preheating facilities	130 or Legal		1.00		0.10						150	40		32.1	(81)	
6. An oil for use in burners equipped with preheaters permitting a high viscosity fuel	150 or Legal		2.00[f]								300	45			(638)	(92)
PS 300	150 or Legal		1.0								40	25				
PS 400	150		2.0[f]									60				

[a] Recognizing the necessity for low sulfur fuel oils used in connection with heat-treatment, non-ferrous metal, glass and ceramic furnaces and other special uses, a sulfur requirement may be specified as 0.5% for No. 1, 1.0% for No. 2, and no limit for Nos. 4, 5 and 6.

Other sulfur limits may be specified only by mutual agreement between the buyer and seller.

[b] It is the intent of these classifications that failure to meet any requirement of a given grade does not automatically place an oil in the next lower grade unless in fact it meets all requirements of the lower grade.

[c] Lower or higher pour points may be specified whenever required by conditions of storage or use. However, these specifications shall not require a pour point lower than 0 F under any condition.

[d] No. 1 oil shall pass test for corrosion made in accordance with paragraph 15 *ASTM Specifications for Fuel Oils*, D 396 – 48.

[e] The 10 percent point may be specified at 440 F maximum for use in other than atomizing burners.

[f] The amount of water by distillation, plus the sediment by extraction, shall not exceed 2.00 percent. The amount of sediment by extraction, shall not exceed 0.50 percent. A deduction in quantity shall be made for all water and sediment in excess of 1.0 percent.

PS 300 and PS 400 are Pacific Specification numbers. Balance of table from *Commercial Standard* CS 12-48 and *ASTM Materials Specification for Fuel Oils* D 396 – 48.

PS 300 is a residual oil for use without preheating in furnaces and burners requiring a low viscosity fuel, and commonly described as light fuel oil, domestic fuel oil, or low viscosity fuel oil.

PS 400 is a residual oil for use in furnaces and burners equipped with preheaters permitting a high viscosity fuel, and commonly described as industrial fuel oil, heavy fuel oil, or high viscosity fuel oil.

Source: Reprinted with permission from American Society of Heating, Refrigerating, and Air-Conditioning Engineers, Inc., *ASHRAE Handbook, 1977 Fundamentals* (New York, 1977).

Heat Value.   Fuel oil comes in different grades, designated by numbers from 1 to 6. The smaller numbers relate to oil that has a lower viscosity, as well as other characteristics. See Table 8-1 for a table of specifications for the various grades of fuel oil.

Each of the different grades of oil has a different heat value, or Btu per gallon rating. The approximate value of the different grades is shown in Table 8-2.

If the combustion is complete in the firing process, which means that all of the fuel and air are combined in the process, then the combustion efficiency is 100 percent; if any of the hydrogen or carbon in the oil is not burned in the process, efficiency is something less than 100 percent. The burner and the combustion chamber are designed to work together. If the burner gets out of adjustment, then the balance between it and the combustion chamber is upset, and the efficiency is affected. Because an oil burner is a complex device, keeping it properly adjusted is not easy. The amount of combustion air used in the firing process also affects the efficiency of the burner. It is necessary to provide between 20 percent and 50 percent more air to the burner than that needed for combustion, as a safety precaution. The excess air passes through the combustion process unchanged and goes through the heat exchanger with the products of combustion. The presence of this excess air reduces the efficiency of the exchanger.

TABLE 8–2
Typical Gravity and Heating Value of Standard Grades of Fuel Oil

Grade No.	Gravity API	Weight lb/gal	Heating Value Btu/gal
1	38–45	6.950–6.675	137,000–132,900
2	30–38	7.296–6.960	141,800–137,000
4	20–28	7.787–7.396	148,100–143,100
5L	17–22	7.940–7.686	150,000–146,800
5H	14–18	8.080–7.890	152,000–149,400
6	8–15	8.448–8.053	155,900–151,300

Source: Reprinted with permission from American Society of Heating, Refrigerating and Air-Conditioning Engineers, Inc., *ASHRAE Handbook, 1977 Fundamentals* (New York, 1977).

The best efficiency that can be expected from an older model oil-fired unit is 80 percent. Manufacturers are continually improving the burners and heat exchangers, so higher efficiencies may be possible on newer models.

Unit Sizes. Number 2 burner oil, which is used for fuel in most smaller furnaces, is nominally rated at 140,000 Btu/h per gallon. The smallest residential furnace usually is designed to use a 0.65 gph nozzle, and so the input of this furnace would be

$$140{,}000 \text{ Btu/gal} \times 0.65 \text{ gph} = 91{,}000 \text{ Btu/h}$$

This furnace would be rated as a 91,000 Btu/h unit, as a measure of input. The nominal output of the unit would be 80 percent of the input, or:

$$91{,}000 \times 0.80 = 72{,}800$$

The amount of heat available for heating a building is 72,800 Btu/h.

Residential oil furnaces are available in capacities ranging from about 75,000 Btu/h input to 300,000 Btu/h. Commercial and industrial models are available in much larger sizes.

Controls

Controls used on an oil-fired furnace include a thermostat, limit controls, blower control, and primary control. Several control functions that are normal to other types of heating units are included in the primary control used on an oil furnace.

Thermostat. An oil furnace thermostat is basically a temperature-actuated electrical switch. It contains a bi-metal sensing element that opens or closes a set of electrical contacts. The sensing element and switch are mounted on a movable plate that is attached to an adjustment lever. The adjustment lever permits selection of a setpoint at which the switch opens and closes. The electrical switch is connected to the primary control on a furnace. Inside the primary control the thermostat circuit opens or closes the contacts in a relay that controls the burner on the furnace.

Limit Control. A typical limit control on an oil-fired furnace is a line voltage limit located in the hot electrical power line to the primary control. The limit control is a temperature-actuated electrical switch. A bi-metal sensing element opens a set of electrical contacts if the temperature goes above a certain setpoint. The sensor for the limit control is inserted in the furnace where it senses air temperature either by the heat exchanger or in the warm air plenum. If the limit control opens on a temperature rise above its setpoint, power is disconnected to the primary control, and the unit is shut down.

Blower Control. A blower control is used on a typical oil-fired furnace to control the blower motor. A typical blower control has a bi-metal element that opens and closes a set of electrical contacts on change of temperature. The control is adjusted to close the contacts when the temperature surrounding the bi-metal reaches a predetermined setpoint between normal room temperature and about 140° F. The bi-metal is inserted in a furnace into the air passing over the heat exchanger or in the hot air plenum.

Combustion Safety Controls

The combustion safety controls on an oil furnace require proving of the flame, or the burner will not stay on. Proving means that the flame must be established on a call for heat.

Stack Switch Safety. On older models of oil-fired furnaces, flame proving usually is accomplished by a thermostat-type device placed in the flue outlet of the furnace. This device is called a stack switch. Figure 8-7 illustrates a stack switch installed in a flue.

A stack switch has a bimetal sensing element that goes in the flue. Changes in temperature of the flue gases cause the bimetal element to expand or contract. A mechanical connecting link from the bimetal opens or closes an electric switch in a control box that controls the burner.

Some stack switches are self-contained with a control box mounted on the flue with the switch. Others have the control box external to the actual switch, and the two are connected electrically. On a furnace using a stack switch as a combustion safety control, the temperature rise of the

Stack switch*

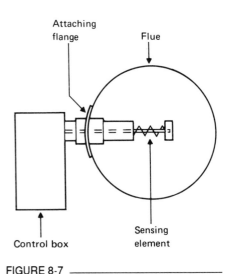

FIGURE 8-7 —————————————————
Stack switch installed in a furnace flue

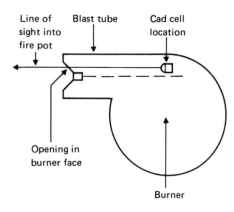

Line of sight into fire pot

Blast tube

Cad cell location

Opening in burner face

Burner

FIGURE 8-8 _____
Cad cell control sensor installed in the blast tube

Cad cell*

Manual reset*

flue gases is monitored by the switch on a call for heat. If the temperature of the flue gases does not go up to a predetermined temperature within a certain time after the call, then the control shuts the unit down on safety lock-out.

Cad Cell Safety Control.   In most oil-fired furnaces manufactured since about 1973, the safety device is a flame scanning unit. The most common one is a **cad cell** unit. The unit uses a cell with a chemical, cadmium sulfide, on the sensing surface. This chemical is sensitive to light in the part of the spectrum related to flame and fire. The cell is placed in a position in the burner so it can "see" into the combustion chamber. Electrical leads from the cell to a control center provide an electrical circuit between the two. Figure 8-8 shows a cad cell located in the blast tube of an oil burner. On a call for heat, if the cad cell does not "see" fire, it will lock out the burner through the control center.

The cad cell control reacts more quickly than the stack switch to fire outage, since the light-sensitive cell and the electronic controls act almost instantaneously. A stack switch works with a bi-metal switch, which is much slower. The cad cell system primary control locks out the burner through a series of relays and controls, just as the primary control does on a stack switch. Figure 8-9 shows a primary control used with a cad cell; it also shows the wiring diagram of the control.

Reset.   One important feature of any of the combustion safety systems is the requirement that the safety have a **manual reset** in case of a lockout. In this case, the control must be physically reset. The primary control or stack switch control has a reset button or lever. This feature is included so the unit can be checked for the cause of failure in case of a lockout. A safety control should never be reset continuously on a failure fire. Each time the control is reset, a quantity of oil is introduced into the combustion chamber. If the unit is not firing, a buildup of oil can occur and a dangerous situation created.

8.3 SUMMARY

Although all of the furnaces commonly used to heat residential or smaller commercial buildings look alike, there

FIGURE 8-9
Control center used with cad cell for combustion safety and wiring diagram of the center

are some major differences in the burners, heat exchangers, and the combustion safety controls used for each of the different fuels.

Oil-fired furnaces basically use a power type burner as a source of heat. Most oil-fired furnaces have a drum-type heat exchanger; and the combustion safety controls sense main fire ignition, rather than prove a pilot light.

In an oil burner, the oil is partially vaporized as it leaves the nozzle in the burner face, and combustion air is mixed with it there. Ignition is caused by an electric spark at the burner face. The power burner produces a large flame that burns in a combustion chamber, which is part of the heat exchanger.

The products of combustion pass up through the heat exchanger proper, which is located above the combustion chamber, and air passing around the outside of the exchanger is heated by the exchanger.

The combustion safety controls are one of two types. A stack switch senses temperature rise in the flue on ignition; a cad cell control actually sees the flame. The stack switch generally is found on older units, and the cad cell safety is used on more of the newer models. The combustion safety controls are used to prevent flooding of the combustion chamber with oil before ignition, which would create a highly dangerous condition.

8.4 TERMS TO KNOW

pot-type burner
atomization
burner nozzle
oil pump
combustion assembly
blast tube
burner face
electrodes
ignition assembly
combustion air blower
gun assembly
drum
stack switch
stack switch safety
cad cell
manual reset

9.1 INTRODUCTION

An electric furnace is one in which air is heated by electrical resistance coils, and since there is no combustion, the furnace does not require a flue or chimney. Otherwise, an electric furnace is similar to a combustion unit. The controls on an electric furnace usually are more complex than those on a gas or an oil furnace. In this chapter, a general description of a typical electric furnace is given, and a specific description of the heating elements and controls are covered.

9.2 MAJOR PARTS

The general appearance of an electric furnace is similar to that of a gas or an oil furnace. No fire is needed in the operation of an electric furnace, so it does not require combustion air or a flue to exhaust the products of combustion. Electric furnaces are available in upflow, downflow, and horizontal air flow. They also are available in a variety of types, such as self-contained furnaces, unit heaters, and duct heaters. Figure 9-1 shows a typical electric furnace.

Many of the parts found on an electric furnace are the same as those in combustion furnaces: filters, blowers, and some of the controls. Figure 9-2 is a cutaway view of an electric blower coil unit showing the internal arrangement of parts.

Parts that are the same as those in the other types of furnaces are discussed in other chapters and will not be repeated here. Instead, this chapter focuses on the parts that are unique to electric furnaces: the heating elements and the controls.

Heating Elements

Electric heating units do not have burners and heat exchangers like combustion-type furnaces. They produce heat through **electric resistance coils.** These resistance coils generally are one of two kinds: (1) nichrome wire or (2) Calrod.

Nichrome Elements. **Nichrome elements** are made with resistance wire of a nickel and chromium alloy. This com-

Electric resistance coils – Wire coils made of material with a high resistance to electric current flow. (When current passes through the wire, the wire becomes hot.)

Nichrome element*

*Term defined in text.

FIGURE 9-1 ————————————————
Typical electric furnace used for residential or
light commercial buildings

FIGURE 9-2 ————————————————
Electric heating blower coil unit showing the
internal parts

Calrod*

Watt (W) – An electrical term used to
describe electrical power. (Technically, it
is the amount of electrical power being
used when 1 volt of electrical potential
moves 1 amp of current; 1 watt of electrical power is equivalent to 3.416 Btu/h of
heat energy.)

Kilowatt*

bination of metals provides a wire with the proper resistance to electrical flow, and with the tensile strength when
heated, to make a good heater. Coils of the wire are suspended in a metal frame so the air can pass readily through
them. The wire is insulated from the frame to prevent an
electrical short circuit. Figure 9-3 shows such an element.

The nichrome wire has a high resistance to the flow of
electric current. When the wire is energized by an electrical power source, the wire becomes very hot. The air that
is to be heated is blown across the hot resistance elements
and thus is heated itself. The heated air is then used for
heating a building or spaces within a building.

Calrod Elements.   **Calrod** is the brand name for elements
that are constructed with a resistance wire inside a protective tube. The wire runs through the center of the tube, but
is surrounded by an electrical insulating material. Figure
9-4 contains an illustration of the construction of a Calrod
unit.

When electric current passes through the resistance
wire, the wire gets hot. The insulating material and the
outer metal tube also get hot. The air passing over the
outer tube is then heated.

The main difference between the nichrome unit and

FIGURE 9-3 ───────────────
Electric heating element of the exposed-wire type

FIGURE 9-4 ───────────────
Drawing of a Calrod-type electric/resistance heater showing the construction details

the Calrod unit is that the air is not directly exposed to the hot resistance wire in the Calrod unit.

Heating Capacity

The heating capacity of individual furnaces is determined by the number of heating elements used in each furnace. The more heating elements, the greater the heating capacity.

Manufacturers give the heating capacity for the units in Btu/h or **watts (W),** and the output is always given for a specific voltage and phase of power used. In smaller electric furnaces, those used in most residential or small commercial applications, the voltage is 240 V and single phase. In larger electrical furnaces, used for larger commercial applications, the voltage can be higher, and the unit may be single or three phase. Voltage is the electrical force, and phase is related to the alternations per second. Btu/h is a term related to heat energy, and watts is related to electrical energy. When electrical energy is converted to heat energy, 3.416 Btu/h is produced for each watt of electrical energy, the term **kilowatt** (Kw), which is 1000 W, is often used in calculations. Watts can be converted to Btu/h by multiplying W times 3.416, and kilowatts can be converted by multiplying Kw times 3416.

These relationships can be clarified by an example. The heat output of an electric heating unit is found by using the following formula:

Btu/h = 3.416 W

when

Btu/h = heat output in Btu per hour
3.416 = Btu/h per W of electric power used
W = electric power used by the furnace

The wattage rating of the furnace is provided in the manufacturer's catalog.

EXAMPLE 9-1

Determine the heating capacity of a furnace rated at 15 Kw by the manufacturer at 240 V single-phase power (240/1/60).

Solution

Multiply 15 kW times 1000 to get 15,000 W. Then multiply the 15,000 W times 3.416:

Btu/h = 3.416 W
      = 3.416 × 15,000
      = 51,240

Therefore, 51,240 Btu/h is the rated capacity of the furnace.

Nearly all electric heating elements are almost 100 percent efficient. All but about 2 percent of the electrical energy input to the element is converted to heat energy. When designers are selecting an electric furnace, they should be sure that the voltage shown in the manufacturer's literature, for a particular unit, is the same as that available on the job.

Manufacturers make a large number of furnace models of differing heat output capacity, either by using heating elements of varying kW rating or various numbers of elements. The smaller units for residential or small commercial use start with 2 1/2 kW elements and increase in approximately 2 1/2 kW increments to about 30 kW. This range provides a wide selection of capacities, ranging from 8540 Btu/h to 102,480 Btu/h. Commercial and industrial units are available in much larger sizes.

Controls

The unit controls used on an electric furnace usually include a thermostat, sequencer, blower controls, and limit controls. In older models of furnaces most of the controls, if not all of them, were typical electric controls. In most of the newer units the controls are electronic, however, their function is the same.

Thermostat. An electric control system thermostat used for an electric furnace heating system is similar to those used for combustion fired units. It includes a sensing element, that is normally a bi-metal element that opens or closes a switch on temperature change. The sensing element is mounted on a movable base that is connected to a setpoint indicator. As the setpoint is adjusted, the make and break point of the switch also is adjusted. The two, setpoint and make and break point are calibrated to be the same. The switch of the thermostat is wired into the control circuit to the control relay or sequencer of an electric furnace.

Because the control circuit of an electric furnace does not usually have as high a current draw as a gas or oil-fired furnace, when a thermostat that can be used for gas or oil operation is used in an electric furnace control system, a resistor often is wired in parallel with the contacts of the electric furnace control relay to increase the amperage draw for proper operation of the anticipator ciruit. This resistor usually is wired into the circuit inside of the thermostat.

Sequencer. Most electric furnaces have multiple heating elements. To prevent too large an amperage draw from occurring all at once when the furnace is energized, the heating elements are brought on the electric line one at a time. A sequencer usually is used to do this. A sequencer of the type used in most electric furnaces is a series of electric switches that close and open one at a time after the first one is energized. Most small electrical sequencers have an electrical heater coil that is energized by control voltage. As the heater coil warm up it closes switches in sequence as a bi-metal element is warmed by the heater. When the heater is deenergized by the control circuit being opened, the switches open again in sequence. The heater coil in the sequencer is wired into the control circuit from the thermostat.

Primary limit*

Secondary limit*

Blower Control. The blower control in an electric furnace is either a relay that is part of the sequencer, or a separate relay that brings on the blower when the furnace thermostat calls for heat. The blower control is used to bring on the blower either as soon as the first heating element is energized, or after a short time delay. It should also keep the blower on until the last heating element goes off on an off cycle of the unit.

Limit Controls.   **Primary limits** are used to protect against high temperature in case of a blower failure. **Secondary limits** are used to protect against high amperage in the operating circuit due to an electrical short circuit. Figure 9-5 shows the limits found on an electric furnace and a diagram of how they are wired into the circuit.

FIGURE 9-5
Electric heating element showing the limits and wiring diagram showing how the limits are wired into the furnace circuits

The primary limit is a thermostatic device with a bi-metal switch in it. If the temperature of the air around the bi-metal gets too high, the switch will open. The switch is wired into the control circuit so that if it opens, the electric element is deenergized. This limit is mounted on the heating element in such a way that the limit senses the temperature of the air going through the element. In case of a blower failure or any other malfunction that decreases the quantity of air flowing, the temperature of the air rises, and the limit switch opens.

The secondary limit is a fusible link. This limit functions in the same way as a fuse does in any other electric circuit. If an overcurrent condition is sensed, the limit overheats and melts, breaking the circuit. The overcurrent condition generally is caused by a short circuit. The fusible link cannot be reset. If it opens by melting, the element must be replaced after the cause of the short circuit is found.

9.3 SUMMARY

Electric furnaces heat air by passing it through hot electrical heating elements that are energized by electric power. The furnaces also use electricity for operating their blowers and controls. The units are similar in appearance to furnaces that use a combustion process for heating the air, although they do not require a flue.

Electric furnaces are available in upflow, downflow, and horizontal air flow and in a variety of types.

The heat-producing devices in electric furnaces are electric resistance heating elements. Two different types are used: bare nichrome wire elements and Calrod elements. In both types of elements, the heat is produced when a current of electricity flows through a wire with high resistance to current flow, and the wire gets hot because of that resistance.

The heating capacity of individual furnaces is determined by the number of heating elements used in each furnace. Normally the elements are rated in watts (W), or kilowatts (kW). These terms are related to the amount of electric power the unit will use to produce heat. One W of electric power will produce 3.416 Btu/h of heat, and one kW will produce 3416 Btu/h.

Combustion safety controls are not needed on electric furnaces, but limit controls are required to shut off the power in case of malfunction. Two such limits are normally used on each set of elements in a furnace: the primary limit and the secondary limit. The primary limit is a thermostatic device, which opens the control circuit if the temperature of the air around the elements goes too high. This condition would occur if the blower went off or anything else stopped the flow of air through the elements. The secondary limit is a fusible link. This link is installed in the line voltage circuit to the element. In case the circuit becomes overloaded or starts to draw too much current, the link melts and the circuit is opened.

9.4 TERMS TO KNOW

electric resistance coils
nichrome element
Calrod
watt
kilowatt
sequencer
primary limit
secondary limit

10.1 INTRODUCTION

In a warm air heating system, air is used to carry heat from the point of generation to the point of use. A hydronic heating system uses a liquid medium to carry the heat. Water and steam are the two most common mediums. Other mediums are used in special applications. Water is used for most small jobs, including residences and small commercial buildings. Steam is often used for larger buildings. The systems and equipment described in this text are for water systems only.

This chapter gives a description of the total hydronic heating system, as well as the boilers, piping or distribution systems, circulating pumps, terminal units, and controls used in hydronic systems.

10.2 HYDRONIC SYSTEM

A **hydronic heating system** is one in which the medium for carrying the heat from the point of generation to the point of use is a liquid. In most cases, that medium is water. Water is commonly used because it carries a large amount of heat per unit. The specific heat of water is 1.00; 1 Btu of heat is needed to raise the temperature of 1 lb of water 1°F. In other words, 1 lb of water can hold 1 Btu of heat for every 1° temperature that the water is raised. The specific heat of water is higher than that of most other materials, which means that other materials have less heat-holding capacity. Table 1-1 in Chapter 1 shows the specific heat for some common materials.

Hydronic systems are used where having the heat-generating plant in one place is convenient, but the point of use is remote from that place. Transporting heat through a small pipe with water is more economical than transporting it through a large duct with air when the point of use is distant from the point of generation. High-rise and cluster-style buildings, such as campus-style school complexes, are good examples of the types of buildings in which hydronic systems are used.

The major parts of a hydronic system are the boiler in which the water is heated, the piping system to carry the water to its point of use, pumps to move the water through the system, terminal devices to exchange the heat

CHAPTER

10
Hydronic Heating Systems and Equipment

Hydronic heating system*

*Term defined in text.

Terminal units

Boiler

Piping

Circulating pump

FIGURE 10–1
Schematic diagram of a typical hydronic system showing the major parts of the system

Boiler – A closed vessel in which a liquid is heated or vaporized.

Terminal device – The device that extracts heat from the water in a hydronic system. (The device usually also distributes the heat into the spaces to be heated as warm air.)

from the water to air, and controls for the boiler, pumps, and the terminal devices. Figure 10-1 is a diagram showing these major parts.

Heating Plant

The heating plant for a hydronic system is the boiler. The **boiler** produces hot water for the entire system. A boiler may be fueled by any one of the common fuels. Its actual configuration depends on the fuel used, as well as the capacity required and the type of supply system used. Figure 10-2 is an illustration of a typical small boiler and some of the other parts of the system. This type of boiler is used primarily for small systems or small zones in a building.

Distribution System

Hot water generated in a boiler, is distributed to the point of use by a piping system. Any one of several layouts of piping can be used, such as single or double pipe, or series or primary-secondary. These layouts are described in Chapter 19. The actual layout selected depends upon the type of terminal devices used and the temperature control requirements in the building. Distribution systems are described in detail in Chapters 19 and 20.

Terminal Devices

The final step in heating a building with a hydronic system is the use of the hot water from the distribution system to heat air at the point of use. Units called **terminal devices** are used for this purpose. Several types of terminal devices are available. The ones used on a job will be selected to fit the physical and air temperature requirements for that particular job. Terminal devices are covered in detail in Chapters 19 and 20.

10.3 EQUIPMENT

The major pieces of equipment used in a total system are the boiler, the circulating pumps, and the terminal devices.

Boiler

A boiler is a device designed to transfer heat produced by combustion, or electric elements, to water. The term **boiler** is used interchangeable for both the heat exchanger of a boiler package and for the entire unit. Boilers are available in different types of various applications. Only the main types will be covered in this text.

Four general parts make up a boiler package:

1. Cabinet or frame
2. Heat exchanger (boiler)
3. Burner or heating elements
4. Unit controls

Cabinet or Frame.   Most smaller boilers, used for residences or small commercial buildings, are enclosed in a sheet metal cabinet. Larger boilers, used for commercial or industrial applications, are built in a frame. In both cases, the cabinet or frame holds the parts of the unit together. In boilers that are built mainly for use in residential heating systems, the cabinet is designed to be attractive as well as functional.

Heat Exchanger.   Boilers are grouped into classes based on working pressure, fuel or energy used for heat, materials of construction, and/or general arrangement of the tubes containing the water.

Hot water boilers are designed to operate at low or medium pressures, ranging from a low of 15 **pounds per square inch (psi)** to a high of 160 psi. Most water boilers are low-pressure type, designed to operate at no more than 30 psi working pressure.

Boilers are designed to burn wood, coal, fuel gas, or fuel oil. Electric boilers also are commonly used.

Some units are designed to use either one of two fuels. These devices are called dual-fuel units. The units can be switched back and forth from one fuel to the other depending upon fuel availability.

Boilers are constructed with either cast-iron or steel heat exchangers. The material used in a particular boiler depends on the size of the unit and its application.

The arrangement of the tubes in the boiler affects the heating unit's general efficiency. Although many different variations of the arrangements are possible, the two most

FIGURE 10–2
Small boiler used for heat in a hydronic system for a residence or small commercial application

Pounds per square inch (psi) – Force or pressure exerted on a surface in terms of pounds force per square inch area of surface.

common are the **water tube boiler** and the **fire tube boiler.**

In its simplest form, a boiler is a water tank with a burner under it. The heat from the burner heats the water in the tank. Most boilers used in hydronic systems are more complex, but they perform the same function. The efficiency of the boiler can be increased by increasing the surface area of the metal separating the burner from the water. The water either goes through tubes above the burner, or the hot flue gases go through tubes in the water tank. When the water goes through tubes above the burner, the boiler is called a water-tube boiler. When the flue gases go through tubes in the water tank, the boiler is called a fire-tube boiler. Each of these types, as well as the electric boiler, is described in the following sections.

Water-Tube Boiler.   A water-tube boiler is one in which the water is contained in tubes above the burner, or the firebox. The use of tubes in the place of a simple tank enables the efficiency of the boiler to be increased considerably. The more surface area of the water-containing tubes that is exposed to the hot flue gases, the more efficient the heat transfer will be. The tubes are connected to reservoirs, or tanks, on each end of the boiler. The reservoirs and the tubes always contain water. The hot products of combustion pass around the tubes and then leave the boiler at one end through a fitting called the **breeching.** The heat from the products of combustion is conducted through the walls of the tubes into the water. Figure 10-3 is a cutaway drawing showing the construction of a water-tube boiler.

Fire-Tube Boiler.   In the fire-tube boiler, the upper part of the boiler is a water tank. Tubes are connected to end breechings and pass through the water tank. The firebox is located below the water container. The water is heated by radiation and by hot flue gases. The hot flue gases also pass into a breeching at one end of the boiler and then go through tubes to the breeching at the other end. Figure 10-4 is a cutaway drawing of a fire-tube boiler.

Electric Boiler.   Water-tube or fire-tube boilers can be oil or gas fired. Electric boilers are a third type of boiler.

An electric boiler is basically a water tank with electric heating elements inserted in the water. When an electric

Water
tubes

Fire
pot

FIGURE 10–3 _____
Water-tube boiler. The water tubes run through the areas above the fire pot where the hot flue gases can surround them.

Boiler, water tube – A boiler in which the tubes contain water and steam, and the flue gases and flames heat the outside of the tubes.

Boiler, fire tube – A boiler with straight tubes that go through the water container and through which the flue gases pass.

Breeching – Open spaces at the end, or ends, of the boiler where the smoke and products of combustion pass from the boiler tubes to the flue outlet.

FIGURE 10–4 _____
Fire-tube boiler. The tubes running through the
water tank above the firebox increase the area
of the surface exposed to the hot flue gases.

FIGURE 10–5 _____
Electric boiler

current flows through the elements, the water is heated.
The elements are insulated from the water electrically so
an electrical short circuit will not occur. An electric boiler
has fewer parts than a gas or oil-fired unit because no
combustion takes place. Electric boilers are more compact
and smaller than gas or oil-fired units with comparable
heating output. An electric boiler is shown in Figure 10-5.

Burners.   The burners used in fuel-fired boilers are similar
to those used in warm air heating equipment. Both atmospheric and power-type burners are used. And, just as in
warm air equipment, the atmospheric burners usually are
used in smaller units and the power burners almost
always in larger ones. Figure 10-6 is a photograph of a typical oil burner used in a boiler. The general construction of
the burners used in boilers is the same as those described
in Chapters 5 and 6. Readers should refer to those chapters
for details of the burners used in boilers.

FIGURE 10–6 _____
Typical oil burner used in larger boilers

Supply pipe*

Return pipe*

Circulating pump – A pump used to circulate water through a piping system.

Piping System

The piping, or distribution, system in a hydronic heating system is the part that carries hot water from the boiler to the points of use and returns cooler water from that point to the boiler. A **supply pipe** carries the hot water to the various terminal devices, and a **return pipe** brings the water back to the boiler after the water has gone through the devices. In most cases, the pipe is either iron or copper.

Unlike the boiler and other parts of the total system, the piping system is custom designed for each job; each system is unique. For this reason, describing a system specifically is impossible. Heating technicians, however, should be able to recognize and describe some general types of systems. These systems are described by the number of pipes used. In a one-pipe system, the supply and return functions are included in a single loop of pipe that goes around the building or the zone to be heated. In a two-pipe system, the supply water is carried in one pipe, and the return water comes back in another. Three- and four-pipe systems are used when both heating and cooling are being done by the one system. Each of the systems is described in detail in Chapter 19.

Circulating Pumps

Circulating pumps are used in hydronic systems to move the hot water through the piping system from the boiler to the terminal devices. These pumps usually are centrifugal pumps driven by electrical motors. Most smaller pumps are built with the motor frames as part of the pump housing. Larger pumps often are indirect drive. That is, the motor is separate from the pump housing, and the pumps are connected either in-line with couplings or are belt driven.

A water pump has an impeller wheel that turns at a relatively high rate of speed. The impeller wheel has vanes attached to a hub, and as the wheel rotates within its housing, it moves water through the pump. Water is drawn in through a suction port in the center of the wheel housing and is then thrown to the outer perimeter of the wheel by centrifugal force. A discharge port is located on the outer rim of the wheel housing. The suction side of the pump is connected to the return piping of the system, and the dis-

FIGURE 10–7 _____
Floor-mounted circulating pump with the motor mounted external to the
pump and connected with an in-line coupling

charge port is connected so it provides water to the supply
side.

Circulating pumps are built in various configurations.
The two that are used most commonly in hydronic systems are floor- and line-mounted pumps. As the names
imply, the floor-mounted pump sits on the floor or a
stand, and the line-mounted pumps are supported by the
pipes to which they are connected. A floor-mounted pump
with an in-line drive with the motor external to the pump
is shown in Figure 10-7.

In most systems, the circulating pump is located in the
return side of the piping system. This placement facilitates
the plumbing for make-up water to the system. All systems must have a source of make-up water available, usually from the water supply used for the rest of the building. As the system operates, air separates from the water
and is removed. There also may be leaks in various parts
of the system. Make-up water is used to replace any volume lost so the boiler will not go dry. Putting the circulating pump on the return side of the system, where the
make-up water comes in, helps to ensure a supply of water
when it is needed.

System Accessories

Hydronic systems require some accessory devices in the water piping next to the boiler. Among these are an **expansion tank** and a **pressure relief valve.** Proper design of the piping arrangement for supply, return, and make-up piping is extremely important. Improper design or installation of

FIGURE 10–8 _____

Diagram of the plumbing connections at the boiler for a typical heating system showing the accessories necessary for the proper operation of the system this piping can cause air traps to be formed in the system. An air trap will prevent the flow of water, a condition that would cause problems within the system. A systematic arrangement of this part of the system is shown in Figure 10-8.

Expansion Tank.   Most hydronic heating systems are closed systems; that is, they are not open to the atmosphere. Because water expands when it is heated, provisions must be made for that expansion in the system. A small tank that normally is full of air during operation of the system is plumbed into the water system above the boiler. The water in the system expands when it is heated and is forced into the tank. The air in the tank can be compressed because it is a gas. The water in the system is not compressible in the liquid state. On an off cycle, when the water in the system cools, the water in the expansion tank flows back into the system, and the air in the tank expands.

Use of the expansion tank allows the system to operate safely and quietly and prevents problems related to high pressure caused by temperature rise of the water. A typical expansion tank and plumbing for it is shown in Figure 10-9.

Pressure Relief Valve.   Every hydronic system must have certain safety features incorporated in its design to ensure safe operation of the boiler and the total system. One of the most important is a pressure-relief valve that opens if the pressure in the boiler or the piping system goes above a predetermined safe operating pressure. Usually the valve is held closed by either a spring or a weight. If pressure in the system exceeds the set pressure, the valve opens, and water is allowed to flow out of the system until the pressure is reduced to a safe level. The discharge port on the pressure relief valve is piped to a floor drain or other water location. A cutaway view of a pressure relief valve is shown in Figure 10-10. Some pressure relief valves also will open if the temperature of the water exceeds a safe limit.

If a boiler's burner fails to shut down for any reasons after the water has reached a **setpoint temperature,** the temperature and pressure could rise to a dangerous level. If the water temperature rises, so does the pressure. In such a case, the pressure relief valve will open and vent

FIGURE 10–9
Expansion tank used with a hydronic system to allow for the expansion of the heated water

Expansion tank – A tank filled with air used on a hot water system to provide a cushion for the expansion and contraction of the water in the system as the water is heated and cooled.

Pressure relief valve – A valve that automatically opens if the pressure inside the tank it is installed on goes above the setpoint of the valve

Setpoint temperature – The desired temperature at which a thermostat is set.

FIGURE 10–10
Cutaway view of a pressure relief valve showing the spring that holds the valve closed against water pressure up to the setpoint of the valve

Aquastat*

Cycle – Alternating periods of time in which the burner is turned off and on.

Convector – A heating coil enclosed in a cabinet with arrangements made for passing air through the unit to be heated by the coil.

Unit ventilator – A blower-powered convection heating unit with a fresh air intake and damper controls.

the system. This process not only prevents the pressure from getting too high, but the operating personnel are alerted to the situation by the discharged water.

Boiler Plumbing.   The plumbing, or piping connections between the boiler and the other parts of the system, such as the supply and return piping, the make-up water supply, and the expansion tank, can be complex. A typical plumbing setup is shown in Figure 10-8. This piping must be designed and installed so that the proper water level is maintained in the boiler, the distribution system always has a water supply, and there are no air pockets in the system.

To ensure that the piping connections for any given system are correct, the boiler manufacturer's recommendations should be closely followed. Proper valves and controls should be installed according to the manufacturer's instructions.

Controls

The control system for a hydronic heating system has two separate parts. The first part includes the controls that regulate the water temperature in the boiler. The second part includes the controls that operate the terminal devices where the heat from the water is converted to warm air for heating the spaces.

Boiler Controls.   The main operating control for boiler water temperature is called an **aquastat**. This control is a thermostat that senses the boiler water temperature and triggers the **cycle** of the burner to maintain the temperature as determined by the setpoint. The aquastat is usually a remote-bulb thermostat, meaning that the sensing bulb is located where it will sense the water temperature, while the operating part of the control is located outside the water container. Most aquastats are temperature-pressure controls. A material in the remote bulb senses the temperature. This material exerts pressure on a diaphragm that operates a switch. If the temperature goes up on the bulb, then the pressure goes up, and the diaphragm is depressed. If the temperature on the bulb goes down, then the pressure in the bulb goes down, and the diaphragm is released.

Since the aquastat controls the temperature of the heating medium, water, this device functions as both an operating control and a limit control. The upper limit of the water temperature, as well as the lower limit, is controlled by the aquastat.

Terminal Device Controls. The second part of the control system is on the terminal devices, the units that transfer the heat from the water to air. In the case of radiators or convection units, the components of the control system usually are very simple. A room thermostat operates a water valve; this valve controls the flow of water through the radiator or coil in the **convector.** The valve is motor operated. Figure 10-11 illustrates a motor-operated valve. The motor is mounted on the top of the valve body and is connected directly to the valve stem. When the thermostat calls for heat, the motor turns, and the valve is opened. When the thermostat is satisfied, the valve motor turns the opposite way, closing the valve. Sophisticated controls are not needed on a simple system because the response time of the system is very slow. The large amount of water in the system causes a time lag between the call for heat and the increase of temperature in the water to satisfy that call.

In the case of blower coil units, the thermostat may control the water flow by the use of a motorized valve, or it may control the blower in the terminal device to control the blower in the room temperature. In more sophisticated systems, both the water and the air may be controlled. Circulated air systems have a shorter response time than do radiation systems on a call for heat.

Whenever outside air is introduced into any air system ahead of a water coil, as in a **unit ventilator,** low-temperature limit controls must be used in the airstream to make sure that the air temperature will not go below the freezing temperature of the water in the system. This control is in the form of a low-temperature thermostat. This thermostat closes the outside air dampers if the temperature of the air entering the coil falls below 32°F, thus protecting the water in the hydronic system from freezing. As an added protection, some systems have a flow control device in the water piping. If the device senses that the water flow has stopped, then the damper is closed.

FIGURE 10-11 _____
Motor-driven control valve for a hydronic system

10.4 SUMMARY

A hydronic heating system is one that uses a liquid-water or steam to carry heat from the point of generation to the point of use. Water is the liquid used in most smaller buildings and is the only one considered in this text.

The major parts of a hydronic system are the boiler, piping, pumps, terminal devices, boiler accessories, and controls. Boilers are self-contained units; the burners and controls for operation are part of the basic package. The accessories to the boiler and piping system, such as circulating pumps, expansion tank, pressure relief valve, and other controls and valves, are selected for each job individually. Each part of the total system has to be matched to all other parts to ensure a properly operating system.

Boilers can be categorized in several different ways. The most important are by fuel or energy used and by the general type according to the way the water is heated. Boilers are manufactured to burn any of the common fuels, and/or to use electricity for heat. Two basic types are related to how the water is heated.

In a water-tube boiler, the water is contained in tubes located over the burner. The use of tubes makes it possible to increase the area of the transfer surface, or the surface exposed to the hot flue gases. A fire-tube boiler has tubes going through the tank containing water, and the hot flue gases go through the tubes. Electric boilers have electric resistance heating elements in water.

The burners used in boilers are similar to those used in warm-air heating units. Both atmospheric and power burners are used. Burner selection depends upon the fuel used and the size of the boiler.

A hydronic heating control system is made up of two separate sets of controls. One set maintains the temperature of the water in the boiler, and the other regulates the amount of heat given up by the terminal devices used to heat the building. The boiler water temperature is controlled by an aquastat, or thermostat, that senses the temperature of the water and cycles the burner as required. The room temperature is controlled by thermostats in the room. The thermostats operate water valves in the piping system or cycle the blower on the terminal device.

The main safety controls are the pressure relief valve, freeze protection controls, and the high limit function of the aquastat. The pressure relief valve is placed in the piping system adjacent to the boiler and is used to prevent the pressure in the system from exceeding a predetermined limit. Freeze controls, located on any blower unit that introduces outside air, limit the low temperature of the air going across the water coil in the system.

10.5 TERMS TO KNOW

hydronic heating system
boiler
pounds per square inch (psi)
boiler, water tube
boiler, fire tube
boiler, electric
terminal devices
breeching
supply pipe
return pipe
piping system
circulating pump
expansion tank
pressure relief valve
setpoint temperature
aquastat
cycle
convector
unit ventilator

11.1 INTRODUCTION

Unit controls are used to cycle a heating unit on and off to provide the heat needed to maintain a comfortable temperature in a building, and to make sure it works safely. The heat loss of a building is directly related to temperature outside the building. The function of a heating unit is to provide enough heat to maintain a comfortable inside temperature regardless of the outside temperature. Since the outside temperature constantly changes, the heat loss also changes. Therefore, the heat output of a heating unit used to heat a building must change to maintain a constant temperature in the building. Because the instantaneous heating output of most heating units cannot be changed, the unit is cycled on and off to provide the required amount of heat over a given period of time.

Unit controls are those that are located in the heating unit and control the operation of the unit itself. Controls that operate the burner or its elements and those that operate the blowers or pumps normally are considered unit controls. Some safety controls, such as limit switches and overloads, also are unit controls.

In this chapter, the unit controls are divided into three categories: (1) operating controls, (2) safety controls, and (3) combustion safety controls. Each category is explained, and the devices used in each category are discussed.

11.2 UNIT CONTROLS

The proper and safe operation of a heating unit depends upon the controls used to operate it just as much as it does upon the mechanical parts. Those parts of the unit related to fuel control, combustion, and air or water distribution all have to be controlled for proper operation of a heating system. The very nature of the heating process requires that safety controls be used.

The controls used on heating units can be separated into three categories: (1) operating, (2) safety, and (3) combustion safety. Figure 11-1 is a wiring diagram for a typical gas furnace. The controls used are labeled and marked to show the category in which they belong.

The controls in these three categories are directly related to the operation of the units. Other control devices may

Unit controls*

*Term defined in text.

FIGURE 11–1
Wiring diagram for a gas-fired furnace. Controls used for different categories of control are included.

be part of the total control system; they are covered in Chapter 12.

Operating Controls

Operating controls are controls that operate various parts of a heating unit. The controls include a thermostat or operator, valve operator, and possibly other devices that regulate the flow of fuel, steam, hot water, or electricity to produce heat. Most heating systems also have controls that activate the blower or pumps to circulate the heating medium when the burner is on. These devices also are categorized as operating controls. Some operating controls are described in the following sections.

Thermostat. A thermostat is a control device used to turn a heating unit on and off in response to temperature changes. The two most commonly used thermostats are those for electric control systems and those for electronic control systems.

Thermostats used in **electric control systems** include a sensor that responds to temperature changes by mechanical changes. One of the most common sensors used is a bi-metal element. A bi-metal element is made of two metals with different coefficients of expansion bonded together. The element bends back and forth in response to temperature changes. It will bend one direction with temperature rise and another with temperature fall. The element is connected mechanically to a set of electrical contacts. In a heating thermostat the bi-metal element opens the contacts on temperature rise and closes them on temperature fall.

The sensing element and contacts are arranged so the contacts can be adjusted to open and close at a given temperature. This is called the setpoint. An adjustment lever is used to set the thermostat to the setpoint. The contacts are arranged so there is about 2°F to 4°F difference between the on and off temperatures.

Thermostats used in **electronic control systems** normally use a thermistor for a sensing element. A thermistor is a semiconductor through which electrical resistance varies in proportion to the surrounding temperature. In an electronic control system the thermostat must be matched with a central control system that originates a low-voltage electrical signal that is used for a control medium. Low-voltage output from the control panel is circuited to the thermostat, and through the thermistor sensor. The resistance in the thermistor modifies the electrical signal relative to temperature, and a returning signal to the control panel calls for heat or not, as the condition requires. Relays in the control panel control the furnace operation.

Blower Control.   A blower control is used in circulating air heating systems to move air through the heat exchanger of the heating unit and the duct system in a building. A typical blower control used in an electrical control system is an electrical switch activated by temperature. A typical blower has a bi-metal element that opens and closes a set of contacts in response to temperature changes. The blower control is installed on a furnace so the bi-metal element extends back into the heat exchanger area or in the supply air plenum of the furnace. The contacts are in a control box on the outside of the warm air space. The bi-metal element on a blower control closes the contacts when the temperature surrounding it goes above a setpoint temperature and opens the contacts when the temperature goes below that temperature.

The contacts on a blower control are wired in series in the wiring to the blower motor in a heating unit. On a call for heat from the thermostat the heating unit is energized and the air in the heat exchanger area or warm air plenum start to heat. When the air temperature reaches the setpoint temperature of the blower control the blower is turned on. The delay between the call for heat and the blower on time allows the air to warm up so cold air is not blown into the building spaces being heated.

Relays.   A relay is an electrical switch in which one electrical circuit can be controlled by another circuit. Relays often are used as control devices when a low-voltage electrical circuit is used to control a line voltage circuit for operating a motor or other electrical device.

A typical electrical relay contains an electromagnetic coil that is connected to the controlling circuit. The coil controls a set of electrical contacts that are wired into the controlled circuit. An electrical switch in the controlling circuit is used to energize the relay coil. This either closes or opens the contacts. Relays are made in many different styles and types, with contacts arranged to control one or more circuits in several combinations of on and off.

Solenoid. A solenoid, like the coil in an electrical relay, can be used to activate a set of electrical contacts for controlling one electrical circuit with another. Solenoids also are used to operate valves and other control devices.

A solenoid is an electromagnet with a hollow core. Inside the coil, in the hollow space, a movable core is located. The movable core is connected to a flapper. The flapper moves a contact back and forth when the solenoid coil is energized or deenergized electrically. The contact on the flapper makes or breaks a circuit with a fixed contact. The coil of the solenoid is connected electrically in a control circuit, and the contacts are connected electrically in a line circuit controlling a piece of equipment.

Safety Controls

Limit controls, air-flow switches, thermal overloads, and other devices fall into the category of safety controls. These controls are designed to protect the equipment and the occupants of a building in case a unit malfunctions. Some of the safety controls are described here.

Limit Controls. A **limit control** is a temperature actuated electrical switch. A typical limit control contains a bi-metal element that extends into the heat exchanger area or warm air plenum of a heating unit. The bi-metal element actuates an electrical switch. As the bi-metal bends in relation to temperature changes, the switch opens or closes. The switch opens on temperature rise and closes on temperature fall. A typical limit switch is calibrated so the switch will open between 180°F to 200°F. If the temperature in the area where the limit control is mounted goes above the setpoint temperature of the control the switch opens. A limit control is wired with the switch in series in the control circuit that controls a heating unit. If the switch opens the heating unit is shut off.

Motor Overloads. Motor overloads are electrical control devices used to open an electrical circuit in case of overheating or overcurrent flow to an electrical motor. Overloads come in two main types and many variations of each. The main types are those that sense temperature rise and those that sense electrical current flow.

An overload that senses temperature rise has a bi-metal element in it that is positioned to sense the temperature of

Limit control – A temperature-actuated switch that opens on temperature rise. (The switch is used to open the electric circuit to the burner control when a dangerously high temperature occurs.)

the surface on which it is mounted. If the temperature of that surface goes above a predetermined high temperature then the bi-metal element opens a set of electrical contacts. The electrical contacts are wired into either the control circuit or the line circuit to a motor the overload is protecting.

A motor overload that senses electrical current is similar to the one that senses temperature but it has an electrical resistance heater in it that is wired in the motor circuit so it will heat up if the motor begins to draw excessive amperage. If the heater heats up it opens a set of contacts that are controlled by a bi-metal element. The contacts are wired into either the control circuit or the line voltage circuit to the motor the overload is protecting.

Combustion Safety Controls

Combustion safety controls are used to prevent the operation of a unit in case of a dangerous or hazardous situation. If the pilot light on a gas-fired heating unit is not lighted on a call for heat, and the gas valve opens a gas-air mix would flood the combustion chamber, and a spark or flame could cause an explosion. On a typical gas-fired furnace with a standing pilot, an automatic pilot safety is used to monitor the pilot flame. If the pilot is not lighted, the pilot safety control prevents the gas valve from opening. Figure 11-2 is a cutaway view of a pilot safety gas valve and the thermocouple used to monitor the flame.

On a gas-fired furnace with electric ignition a spark igniter lights the pilot light on a call for heat. A thermocouple or other sensing device is located where the pilot light heats it. Some of the other sensing devices used are flame rectification or mercury pressure units. Whatever sensing device is used, it is connected electrically to an electronic control panel in which relays operate to shut down the unit if the pilot light does not stay on, or if it goes out during a firing cycle.

An oil-fired furnace has no standing pilot, so the combustion safety control monitors the start-up of the unit to ensure firing. The control also continuously monitors the combustion during an on cycle to make sure the flame does not go out. The monitoring is done by a stack switch or a cad cell control.

Electric furnaces do not have combustion safety controls, but they do have limit controls on the electric power circuits to the elements. These limit controls serve the same purpose as combustion controls: they monitor the

FIGURE 11–2
Pilot-safety valve and thermocouple used to monitor the operation of a pilot ignition system

FIGURE 11-3

Primary and secondary limits in the wiring on an electric furnace

amperage in the circuit, or the temperature of the air around the element, and shut the heat off if a dangerous situation is sensed. Each electrical circuit to a heating element has two limits. One is sensitive to electrical current flow and the other to temperature. A wiring diagram of a typical electric furnace is provided in Figure 11-3. The limits are shown wired into the circuits. If either limit senses a dangerous condition, the limit opens, and the element will not be energized.

11.3 CONTROLS FOR DIFFERENT FUELS AND ENERGY

The same basic control scheme is used for most types of heating units. The burner, or heating elements, are cycled on and off to provide a specific amount of heat over a given length of time. A controller, such as a thermostat, calls for heat when heat is needed and shuts off the burner or deenergizes the heating elements when the temperature is satisfied. Other controls provide safety functions in case of malfunctions.

These basic control functions are the same for furnaces using different fuels or electricity for the production of heat, but the application of the controls to each type of unit

FIGURE 11-4
Low-voltage wiring from the transformer to the
gas valve through the thermostat

FIGURE 11-5
Combination fan and limit switch used to monitor temperature in the heat exchanger and also
to operate the blower

may vary somewhat. The particular application usually is
determined by the burner or element characteristic.

Gas Furnace Controls

A typical gas furnace control system is a 24 V electric system. A transformer reduces line voltage to 24 V for control
power. The operating control circuit runs from the control
transformer, through a thermostat, to the gas valve and
back to the transformer. The low-voltage line from the gas
valve back to the transformer usually contains the limit
control contacts. See Figure 11-4 for a wiring diagram of
this circuit.

In some larger units, the control circuit may be a line
voltage circuit. The circuit in this case is the same except
that a transformer is not used, and the thermostat and gas
valve are rated for the higher voltage.

The blower control also is part of the operating control
system. A blower control is an electrical switch similar to a
thermostat. The device has a set of electric contacts that
open or close on temperature rise or fall. Figure 11-5
shows a combination fan and limit control. The fan side of
this control operates the blower. The blower control is
located with access to it in the front vestibule of the furnace. The blower control sensing element extends back
into the heat exchanger section of the furnace. When the
burner comes on and heat is produced, the sensing element is heated up. This closes the contacts in the blower

FIGURE 11-6

Diagram showing the blower control wired into the circuit to the blower motor

control. This blower control switch is wired in series with the blower motor. So, electric power goes to the motor through the switch. Figure 11-6 is a wiring diagram of the blower circuit on a typical furnace.

A heating unit also has safety controls that shut the heat source off in case of a unit malfunction. For example, if a gas valve sticks open, or a blower motor fails, the temperature in the furnace could exceed that desired for safe operation of the unit. A limit control wired in series with the gas valve in the operating control circuit prevents this occurrence. Figure 11-7 shows a limit control and a diagram of how it is wired into a system.

A limit control is a temperature control switch that opens if the temperature reaches the setpoint of the switch. The setpoint is a temperature between about 160°F and 200°F, the setpoint being the highest temperature at which the unit could operate safely. On downflow or horizontal furnaces, two limits are often used. One is located in the heat exchanger, and the other is located in the unit where the air leaves the blower. The extra limit is needed in these two types of furnaces because on initial firing, the air may move by gravity in the direction opposite to the way it will when the blower comes on. In this case, the second limit will protect the unit against overheating. The second limit also gives protection in case of blower failure. When two limits are used, they are both wired in series in the operat-

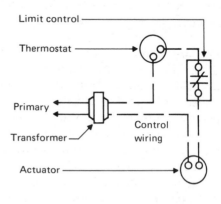

FIGURE 11-7

Limit switch and wiring diagram for a typical furnace control system ing control circuit, so either one can shut the burner down in case of problems. A complete wiring diagram of a typical gas-fired furnace is shown in Figure 11-8.

Oil Furnace Controls

Most oil furnaces, like gas furnaces, have 24 V control systems. The **transformer,** used to reduce the line voltage to 24 V is part of a control package called the **primary control.** A primary control used on an oil furnace is shown in Figure 11-9. The primary control also contains relays and other devices necessary to make the combustion safety circuit. A diagram of the wiring, relays, and switches found in a typical primary control are shown in Figure 11-10.

The primary control regulates the operation of all of the electrical components of an oil furnace. Electrical power is brought directly to the black and white **leads** on the **primary control** box from the **electrical disconnect.** An orange lead goes from the primary control to the oil burner assembly. A wire from the **neutral** terminal on the disconnect is connected to the white lead on the oil burner assembly. Two low-voltage terminals on the primary control are connected to the thermostat, and two more to the combustion safety device. Figure 11-11 shows the circuit connections.

When an oil furnace thermostat calls for heat, a coil in the primary control is energized. The coil closes a set of contacts that bring on the burner and also activate the combustion safety circuits within the primary control itself.

In addition to the circuits just described, the typical oil furnace has a blower motor circuit that operates through a blower control. The hot leg, or power leg from the power source is circuited through a blower control switch that closes when the temperature of the air in the furnace increases to a predetermined temperature. The other side of the blower control is connected to one terminal on the motor. The other terminal on the motor is connected to the neutral side of the power source. This wiring completes the blower circuit, which is the same as that used on a gas furnace.

In downflow or horizontal furnaces, a **sure-start switch** is sometimes used for blower control. In this type

Transformer – An electrical device that changes the voltage received from an electric power source.

Primary control – A control used on oil-fired furnaces that contains the transformer, operating relays, and combustion safety relays and switches for the operation and protection of the unit.

Leads – Electric connections at a device.

Control box – A relay or primary control that is used to control the operation of a unit.

Disconnect – The electric disconnect used to control the electric power to a unit.

Neutral – The common or neutral power line on an electric circuit.

Sure-start switch – A blower relay that is controlled by an electrical signal as well as by temperature. (It will always close within a certain length of time after the electrical signal is received.)

Pictorial

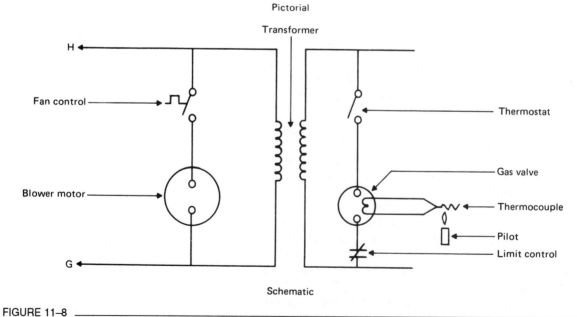

Schematic

FIGURE 11–8
Pictorial and schematic wiring diagrams for a typical gas-fired furnace identifying the control components

FIGURE 11–9 _____
Primary control used to monitor the operation
of an oil furnace

FIGURE 11–10 _____
Diagram of the internal wiring for a typical primary control

FIGURE 11–11 _____
Circuit connection to a primary control

FIGURE 11–12 _____
Wiring diagram for the power to the primary control with the fan and limit control and also an auxiliary limit in series in the circuit

Sequencer*

Time-delay relay – A relay in which the contacts close within a specific length of time after an electric signal is received.

of control, an internal heater is wired in parallel with the thermostat terminals on the primary control so the switch will be activated by the thermostat on a call for heat. Time delay is built into this heater so the air in the system is warmed before the switch starts the blower.

The safety controls on an oil furnace are similar to those described for a gas furnace. One or two limit controls are wired in series in the power line to the primary control. See Figure 11-12 for a diagram of this wiring. If either of the limit controls opens because of high temperature, the unit will shut down.

Combustion safety controls are described in Chapter 8, on oil furnaces. Figure 11-13 shows a complete wiring diagram for a typical oil-fired furnace, with controls identified.

Electric Furnace Controls

The control system used on electric furnaces typically is a 24 V electrical system. The 24 V power source is a transformer. In the operating control circuit, one leg of the transformer is connected to the thermostat. The circuit goes through the thermostat and then to one terminal on a control relay coil, the magnetic coil that operates the relay. The wire from the other side of the relay coil goes back to the opposite terminal on the transformer. On a unit with low voltage limits, the limit control contacts are wired into one of the above circuits. A wiring diagram of this circuit is shown in Figure 11-14.

In a furnace with this wiring diagram on a call for heat from the thermostat, the coil in the control relay or sequencer is energized. This closes the relay contacts and starts energizing the elements. The elements are energized through an electrical or mechanical **sequencer.** A sequencer is a control device that brings the electric heating elements on in steps, one at a time. The sequencer is a series of **time-delay relays,** electrically interlocked, so that each one controls line voltage to an element and also initiates the timing sequence for the next set of contacts. A sequencer used for this purpose is shown in Figure 11-15.

The elements are sequenced on one at a time rather than all at once so the electrical load on the circuits comes on in increments. One of the most important features of the electric furnace is that the heat can be sequenced on in gradual steps. This sequencing closely approximates modulation and enables heat demand to match output more closely. The heat output of the unit can be varied in degrees by varying the number of elements that are energized. Solid-state controls make this operation relatively simple.

Modulate – To change capacity to meet demand.

FIGURE 11–13
Pictorial and schematic wiring diagram of a typical oil-fired furnace

FIGURE 11–14 _____
Electric furnace control circuit with the thermostat in series with the sequencer

FIGURE 11–15 _____
Sequencer for electric furnaces

FIGURE 11–16 _____
Wiring diagram for a typical electric furnace with the controls identified

In some cases the electric elements are cycled on and off by an electronic themostat that is sensitive to small temperature variations. A special blower motor is used to control the speed of the blower and thus regulate the volume of air across the electric elements. A multispeed or modulating blower motor is used, with a motor speed control operated by leaving air temperature of the furnace. With this type of control the discharge air temperature remains constant, even though the electric elements are staged on and off.

The blower on an a conventional electric furnace is controlled by a blower relay. The coil on this relay is wired in parallel with the control relay coil. When the thermostat calls for heat, the blower relay starts the blower. If required, the blower relay may be a time-delay relay, so the blower will not start until the air is warm enough to heat the building.

Safety controls are discussed in Chapter 9. Figure 11-16 shows a wiring diagram for a typical electric furnace.

11.4 SUMMARY

The heat loss of a building is directly related to the outside temperature and changes as that temperature changes. The heat output of most heating units, however, is constant while the unit is operating. To match the heat loss of a building, the heating unit is cycled off and on to provide an average amount of heat per hour required to offset the heat loss per hour. The cycling off and on of the unit is done by the unit controls.

The controls used to regulate the heat output of a unit are called operating controls. Other controls, called safety and combustion safety controls, are also used on heating units to ensure safe operation.

The controls used on units that burn different fuels or use electric energy to produce heat are similar to each other, but their applications differ slightly depending upon the burners or the elements used.

11.5 TERMS TO KNOW

unit controls
limit control

relays
solenoid
motor overloads
combustion safety controls
transformer
primary control
leads
control box
electrical disconnect
disconnect
neutral
sure-start switch
sequencer
time-delay relay
modulate

12.1 INTRODUCTION

A **control system** is the complete control package that operates a heating system. The control system tells the heating equipment when to start up and when to shut down. The control system usually includes some devices that are external to the unit, as well as those that are internal. The thermostat or sensor, actuators or valves, connecting wires and/or tubes, and all the accessories that control the temperature, humidity, or any other condition related to comfort in the building are all part of the control system.

There are four general types of control systems: electrical, pneumatic, electronic, and mechanical. Electrical, pneumatic, and electronic systems are described and explained in this chapter. A mechanical control system is one in which control functions are performed manually. A mechanical control system is used so seldom in modern heating systems that it is not discussed in this book.

12.2 COMMON FACTORS IN CONTROL SYSTEMS

All control systems include components that will provide the following functions:

1. Sense conditions such as temperature, humidity, or pressure
2. Transmit data
3. Actuate some device to satisfy a condition
4. Feed back information when the condition is satisfied

Figure 12-1 is a block diagram showing the relationship of these functions in a system.

Each of the four functions can be observed in a thermostat. The thermostat measures room temperature and compares it with a setpoint temperature. The information is **transmitted** by an electrical signal. If the room temperature drops below the setpoint, an electric circuit is completed by the closing of the contacts in the thermostat. The electric signal through the thermostat actuates a gas valve in the furnace to generate heat. When the temperature in the room comes up to the setpoint the electric circuit is broken. The broken circuit is the **feedback signal** that the condition is corrected.

Control system*

Transmit – In a control system, the signal that indicates the comparison between setpoint and the actual condition.

Feedback signal – A signal indicating that the setpoint and actual conditions are nearly the same.

*Term defined in text.

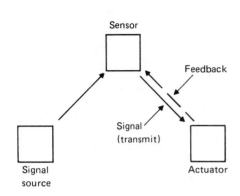

FIGURE 12–1 _____
Block diagram of the four functions of a control
system

FIGURE 12–2 _____
Thermostat, or sensor, used in an electric control system

Electric Control System

In an electric control system, the **sensors** control electric
switches that close or open to allow an electric current to
flow or not to flow. A thermostat in an electronic control
system contains both sensor and a sensor-controlled
switch. Figure 12-2 shows a typical thermostat used in an
electric control system.

A thermostat has a setpoint selector that can be set to the
temperature desired in the space in which the thermostat is
located. If the temperature in the space goes above the setpoint of the thermostat, an internal switch in the thermostat
opens, interrupting the circuit. If the temperature in the
space goes below the setpoint of the thermostat, the internal
switch closes, completing the circuit. Figure 12-3 is a simplified diagram of wiring circuitry in a thermostat.

The thermostat, then, is a switch that controls an electric signal. the signal controls the operation of equipment.

The transmission system for an electric control system,
which transmits the control signals, is the electric circuit.
This transmission system includes the power source and
the wires to carry the signal. The transmission system may
be one circuit or more, but each electric circuit in the control system performs the transmission function.

The **actuator** in an electric control system is a solenoid,

Sensor – A device in a control system
that compares actual conditions with a
setpoint condition and sends a signal
indicating that difference.

Actuator – The device in a control system
that performs a function to correct the
condition causing a difference between
the setpoint and the actual condition.

FIGURE 12–3 _____
Schematic diagram of the internal wiring for an
electric thermostat

FIGURE 12–4 _____
Gas valve, or actuator, used in an electric control system

electric valve motor, or other electrically driven device.
The device opens a valve when a current is flowing in a
circuit and closes the valve when the current does not
flow. The gas valve shown in Figure 12-4 contains a solenoid that controls gas flow. The actuator is wired into the
control circuit in series with the sensor. When the sensor
switch closes, current flows, and the actuator functions.

The feedback function in an electric control system is
normally provided by the sensor. The thermostat opens
when it is satisfied because the temperature has reached
the setpoint. The open circuit is a signal that the need for
heat has been satisfied.

Other controls, such as limit controls, may be wired
into an electrical control circuit to provide safety functions,
but these controls are auxiliary to the operating controls
and not actually part of the basic function of that part of
the system.

Electric controls are basically on–off or digital controls.
The sensor operates an electrical switch. When the sensor
recognizes a need for heat, humidity, or whatever condition the system is controlling, an electrical switch is closed,
and the unit is turned on. When the sensor indicates the
controlled condition is satisfied, the switch is opened, and
the unit is turned off. The actuator used in an on–off control system is also an on-off or digital actuator.

Multistage control systems also are available. In a
multistage control system sensor, there is more than one
set of contacts. Each set of contacts is opened and closed at

Multistage control – A control or control
system that controls the system so the
output of the equipment is varied in
steps or stages.

different control conditions, such as different temperatures. Such a sensor can control multistage burners or separate elements in an electric furnace. Each stage or set of elements is controlled by one of the sets of contacts. Multistage control makes it possible to approximate, more nearly, heat output of the heating equipment to the heat loss of a building.

One other type of electrical control system is used. This is a **modulating control** system. In a modulating control system, a sensor is used that varies the electrical control signal according to the conditions sensed. The varied signal is sent to an actuator that opens or closes a valve or other device in proportion to the strength of the signal received. This control is used with a modulating burner or other device that varies the heating output of the controlled unit to match the load at any condition.

Pneumatic Control System

A **pneumatic control** system provides the same results as an electric control system does. However, in a pneumatic control system the control signal medium is air pressure. The main advantage of a pneumatic control system over an electrical system is that control is inherently modulating in a pneumatic system. The sensors in a pneumatic system modulate air pressure in the transmission lines, and the air pressure is used to position actuators. The position of the actuators is determined by the pressure.

In a pneumatic control system, a compressor is used to compress air and store it in a tank. The air from the tank flows to the system through a filter and a pressure reducing valve. The air leaves the pressure reducing valve at approximately 18 psi. Figure 12-5 is a diagram of a typical pneumatic system.

The sensors in a pneumatic control system provide the same function as thermostats in an electric system, but instead of being on–off electric switches, the sensors, or thermostats, are pressure control devices. The sensors are piped into the system between the air supply tank and the actuator, or operators, as shown in Figure 12-6.

Small-diameter tubes (1/4 in. diameter) bring the air under pressure from the tank to the sensors. The sensor has a small orifice or nozzle through which the air can escape. A movable vane that covers the orifice regulates the escape rate. The movable vane is positioned by a bi-

Modulating control – A control or system that controls the output of the equipment over its operating range, in infinite increments.

Pneumatic control – A control or control system that uses air for the control medium.

Spring return – An activator in which the "on" motion moves against a compression spring. (In the "off" position, the spring returns the actuator to its original position.)

FIGURE 12–5
Diagram of a typical pneumatic control system

metal element in the case of a thermostat, or other sensor element in other controllers. If the temperature surrounding the sensor goes above the setpoint, then the nozzle is closed, and air pressure in the line from the sensor to the actuator increases. If the temperature drops, the nozzle is opened, and air pressure to the actuator is decreased. The diagram in Figure 12-7 shows how the sensor regulates the pressure.

The actuator is a pneumatic piston that is positioned by air pressure. The air leaves the controller and then goes to the actuator. Figure 12-8 shows a variety of actuators. Each has a different action.

The more pressure the sensor allows, the more the actuator is moved open; the less pressure the sensor allows, the more the actuator is allowed to move closed. The actuator is a **spring return** device. It is connected to the valve or other device being controlled. A pneumatically controlled valve is shown in Figure 12-9.

One of the advantages of a pneumatic control system as compared to an electric system is that the pneumatic system is a modulating system. The amount of motion of

FIGURE 12–6
Schematic diagram of the parts of a pneumatic control system showing the relationship between the sensor and the actuator

FIGURE 12–7 _____
Schematic diagram of the operating parts of a pneumatic thermostat

Field controls*

*Term defined in text.

Interface – The point in a system where one type of control system is connected into another type of control system.

the actuator is proportional to the air pressure supplied by the sensor, and the air pressure is proportional to the signal received by a sensor. Pneumatic systems can be used to control heat or humidity, to position damper motors, and to regulate valves. Pneumatic systems usually are used on larger, more sophisticated jobs than are electric systems.

Pneumatic controls usually are used as **field controls** only—that is, the controls outside the unit. In most cases,

FIGURE 12–8 _____
Schematic diagrams of a series of pneumatic operators illustrating the variety of actions possible

FIGURE 12–9 _____
Pneumatic operator mounted on a valve

Adjusting screw

N.O. terminal
N.O. contacts
Common terminal
N.C. contacts
N.C. terminal

Air

Diaphragm

FIGURE 12–10 _____
Diagram of the working parts of a
pneumatic-electric (PE) switch

the controls inside the units are electric, even though the controls external to the units may be pneumatic. When both types of controls are used, a pneumatic-electric (PE) switch is part of the system. This switch changes a pneumatic signal to an electric signal and allows a pneumatic control system to be **interfaced** with an electric one. Figure 12-10 is a diagram showing how a PE switch works.

Electronic Control System

An electronic control system, as the name implies, uses electronic devices to provide the control functions. Normally the electronic devices control electric circuits, and these circuits power the actuators in the system.

One of the major differences between an electronic control system and either the electric or the pneumatic system is the use of a control center. Electronic systems operate on low-voltage **direct current (DC)** power. But

FIGURE 12–11 _____
Electronic control center showing the many electronic devices that operates the unit. The terminals on the right side are used to connect into the operating system.

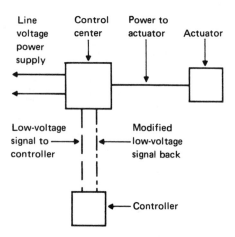

FIGURE 12-12 _____
Block diagram showing the relationship of the main parts of an electronic control system

because high voltage **alternating current (AC)** is the current that normally is available on a typical job, this current has to be transformed down and **rectified** for use in the system. Rectification is the changing of the electrical power from AC to DC in a circuit. Electronic devices perform the rectification. The transformer and rectifier circuit is part of the control center, along with the other devices and circuits. An electronic control center is shown in Figure 12-11.

A typical electronic control center has a rectifier that provides a low-voltage DC signal, usually 0 to 20v or 4 to 20 ma. The signal, used for the transmission function of the system, is circuited to the sensor of the system. The sensor is an electronic device that modifies the signal in some way. The modified signal then comes back to the control center. In the center, this signal then controls a 24 V, or line voltage, circuit that controls the actuator. The sensors in an electronic system are devices that sense temperature, humidity, light, pressure, and other conditions. A block diagram showing the relationship of the parts of an electronic control system is shown in Figure 12-12.

The control center can be designed to receive signals from more than one sensor point. The center can compare the various signals and then send out an actuator circuit signal related to the multiple signals received. For example, if both a room thermostat and an outdoor thermostat are used on a control system, the outdoor thermostat may be used to reset the indoor thermostat setpoint. The indoor thermostat temperature setpoint is automatically raised as the outside temperature gets colder. Electric and pneumatic systems also can perform this function, but not as easily as can an electronic system.

The solid-state device used to sense temperature is a thermistor. The electronically functioning part of a thermistor is a crystal, in which the resistance to an electric current changes as the temperature of the crystal changes. In a typical thermistor used in a heating control system, the resistance increases as the temperature decreases, and the resistance decreases if the temperature increases. When a thermistor is wired in series in a circuit, temperature variations cause variations in the current flow in the circuit. The electrical signal received through the thermistor is sent to the electronic control center, where it is converted to an AC signal of high enough voltage to drive the actuators in the system.

The actuators used in an electronic system are similar in design and function to those used in an electric control system. Figure 12-13 shows a damper motor for use in an electronically controlled system.

The electronic control system combines the low-cost feature of the electric wiring found in a pneumatic system. When those advantages are coupled with the simplistic yet rugged features of the electronic components, it is easy to see why electronic systems are popular.

12.3 SUMMARY

Just as each heating unit in a heating system must have controls for proper operation, a control system is needed to integrate the different parts of a unit, or multiple units, in a building. The control system components are (1) sensors, (2) signal devices, (3) actuators, and (4) feedback devices. These parts tell the heating equipment when to start up and when to shut down. The most common systems are electric, electronic, and pneumatic.

Electric systems use electricity that is carried in normal circuits as a signal. All of the components are electrical; that is, electricity is supplied, controlled, and used as a source of power in each part of the system. Most commonly, electric controls are on–off or digital controls. On–off or digital controls turn devices on, or they turn them off. The controls provide no intermediate positions. Special electric controls, however, provide multiposition control or even modulation.

A pneumatic control system uses air pressure as a signal. Air is compressed and stored in a tank. Small-diameter tubing carries the air to the sensor where the pressure is regulated according to the setpoint of the sensor. The modified pressure is used to position the actuator. In many cases, pneumatic controls are used to control electric signals. Pneumatic controls usually are used on larger buildings and are excellent in cases when modulation is required.

Electronic control systems use variations in DC low voltage or electrical current as signals. Electronic control systems use solid-state devices as sensors; the signal usually is a DC electric signal that is modified by the sensor. The DC signal is changed to an AC signal in a control center. The control center sends the AC signal to actuators in the system. Electronic system components are simple but

FIGURE 12–13 _____
Modulating motor used as an operator in an electronic control system

rugged. System modulation is easily achieved with an electronic control system.

12.4 TERMS TO KNOW

control system
transmit
feedback signal
sensor
actuator
multistage control
modulating control
pneumatic control
spring return
field controls
interfaced
direct current (DC)
alternating current (AC)
rectified

13.1 INTRODUCTION

When the weather is cold, heat is required to keep buildings warm. So far this text has discussed heat and the equipment that supplies it. In this chapter the factors that cause the need for heat in a building are discussed; subsequent chapters explain how to calculate the amount of heat needed, an operation called figuring the heat load. The **heat load** is the total heat, in Btu/h, that must be supplied to keep a building at design temperature. The **heat loss** is the Btu/h lost from the building because of temperature differences.

Traditionally, heat loads for buildings were figured manually, using variables and factors relative to each building separately. Today, most heating and cooling loads are figured on computers, using computer programs that facilitate the many mathematic calculations in the process. An understanding of the variables and factors involved in the calculations is necessary for any technician regardless of how the loads are figured.

To figure a heat load for a building, a designer must determine what causes the load. Causes can be divided into two categories: variables and factors. Heat loss variables include temperature difference between the inside of a building and the outside air temperature. Heat loss factors are numerical values related to the rate of heat conduction through surfaces of a building, and to any cold outside air that infiltrates the building.

An accurate calculation of heat loss is the first step taken when a heating system is designed for any building. All of the decisions concerning the type of system to use, equipment type and size, distribution of air, and controls are dependent upon heat loss. This chapter explains what heat loss is, what causes it, and Chapter 14 explains how to use heat loss to calculate the heat load.

13.2 DEFINING HEAT LOSS

One of the first laws of thermodynamics states that heat always travels from a warm object to a cooler one. An ice cube in a glass of water cools the water because heat from the water is absorbed by the ice. The ice melts because it

CHAPTER

13
Heat Loss Variables and Factors

Heat load*

Heat loss*

Heat gain – The heat in Btu/h that a building accumulates because of the design temperature difference, solar heat, and any other source.

*Terms defined in text.

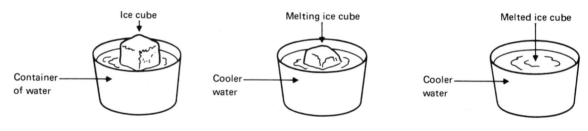

FIGURE 13–1

One law of thermodynamics. Heat always flows from the warmer object to the cooler object, and, in the process, the temperature of each object changes.

Variable – The elements in a heat loss calculation that are generally different for each job.

Factors – Numerical equivalents of the amount of heat that is conducted through a given material in Btu/h per sq ft per degree Fahrenheit.

has absorbed heat, and the water is cooled because it has given up heat. See Figure 13-1.

To relate the melting of ice to the heat loss from a building, consider how the building and the spaces in it are affected by the temperature of the outside air. If the temperature of the air mass surrounding a building is colder than air in the building, then heat will flow from inside the building to the air surrounding it. See Figure 13-2. If the air outside the building is warmer than the air inside, then the heat passes into the building, and **heat gain** occurs.

To maintain a comfortable temperature inside a building when the air temperature outside is lower than the air temperature inside, heat must be provided to the building at the same rate as it is lost. To determine how much heat is needed, a heat loss calculation is made.

The calculation is made by figuring the loss for those parts of the building that heat will pass through, such as walls, windows, ceilings, and floors. These losses are called conduction losses. Losses from other sources, such as air that leaks into the building from outside through cracks, or heat loss because of the air brought in for ventilation, also have to be figured. In all cases, the losses are proportional to the difference in temperature of the air inside and outside the building.

Calculating the Heat Load

In calculating a heat load for any building, it is necessary to identify and consider the **variables** and **factors** that affect the loss in the building. Heat loss variables include temperature difference between the inside and the outside

FIGURE 13–2
Heat loss of a building

of a building, and areas of the building surface that are exposed to outside temperature. Heat loss factors include the rate of heat conduction through the surfaces of the building and the rate of infiltration of cold air into the building. These variables and factors are combined in an orderly format, and the heat loss is figured by following a series of steps.

- Determine the temperature difference between the inside and outside of the building.
- Consider the different types of construction materials used in the building (such as the glass in the windows, the wooden doors, or materials in the walls), and then calculate the areas of each of these.
- Figure the amount of outside air that is leaking into the building through the cracks around the opening parts of the windows and doors, as well as the air that gets into the building when the doors are opened. Figure the fresh air brought in for ventilation.
- Find the appropriate heat transfer factor for each of the building components and for the infiltration by using a factor table.
- Multiply the heat transfer factor for each component by the area of the component, and multiply the heat transfer factor for infiltration by the amount of crackage to determine the heat losses due to conduction and infiltration.
- Add the losses together to find the total heat load of the building.

Terms Used in Determining Heat Loss

Some of the terms used in determining heat loss have been defined earlier in this text; others are new. To help in understanding the procedures explained later in this chapter and throughout the book, a list of the most common terms with definitions and descriptions follows.

Design Temperature.   Design temperature is a selected or arbitrary temperature used in heat-loss calculations.

Outside Design Temperature.   Outside design temperature is the temperature of the outside air for the time at which the heat loss is calculated. Design temperature

Components – The constituent parts of a structure.

Temperature*

Design temperature*

Outside design temperature*

Inside design temperature*

Design temperature difference*

R factor*

U factor*

Temperature – Temperature is the measurement of the intensity of heat. Temperature is measured with a thermometer, which commonly is read in degrees Fahrenheit (standard) or in degrees Celsius (metric).

Btu – Btu is the measurement of a quantity of heat. A Btu is defined as the amount of heat that will raise the temperature of 1 lb of water 1°F. If a common wooden kitchen match is burned, approximately 1 Btu of heat will be generated.

Btu/h – Btu/h relates quantity of heat to a time period. In heat loss calculations, time is a factor, and so the term Btu/h is used to indicate the amount of heat involved in a process during one hour.

tables give the lowest outside temperature that might be expected to occur in the locality for which the designer wants to figure a loss. Table A-3 in the Appendix is a table of outside design temperatures for the United States.

Inside Design Temperature. The **inside design temperature** is the temperature that the designer wishes to maintain in a building. The temperature is an arbitrary one decided upon by the designer. A normal heating design temperature for buildings is between 70°F and 75°F.

Design Temperature Difference. The **design temperature difference** is the difference between the inside design temperature and the outside design temperature. This variable is found by subtracting the outside design temperature. Design temperature difference is the actual temperature difference used in the heat loss calculation.

R Factor. An **R factor** defines the amount of resistance to heat flow through a 1 sq ft area of a given building material for each 1°F temperature difference on each side of the material. The R factor, given in Btu/h, is based on materials as tested in a laboratory. R factors are listed on insulating materials so a purchaser will know the insulating value of the material when it is purchased or used. The higher the number, the better the insulation.

U Factor. U factors are the most commonly used factors for figuring heat loss. A U factor defines the amount of heat, in Btu per hour (Btu/h) that will pass through 1 sq ft of a material for 1°F of temperature difference on each side of the material. U factors normally are given for combinations of construction material such as the assorted building components that make up an outside wall.

U factors are found by dividing the sum of the R factors into the number 1. This operation is called finding the reciprocal. The formula is

$$U = \frac{1}{R_1 + R_2 + R_3 \ldots}$$

The better the insulation value of the components, the smaller the factor.

U factors are used as multipliers in a heat loss calculation. Figure 13-3 is a pictorial representation of a U factor. Tables listing the U factors for various combinations of building components are available from the American Society of Heating, Refrigeration, and Air Conditioning Engineers, Inc.

Heat Transfer Factor.   A **heat transfer** factor (htf) is a factor in which the U factor for given materials has been multiplied by the design temperature difference. To determine heat loss due to conduction, the heat transfer factor is multiplied by the area of exposed surface. The same method is used to determine heat loss due to infiltration. The heat transfer factor may be selected from a factor table, or the multiplier to be used in the heat loss calculation can be obtained by multiplying the U factor by the design temperature difference. When heat transfer factors are selected from a factor table, car must be taken to make sure that they are selected for the design temperature difference of the locality in which the building will be erected. Table A-4 in the Appendix is an example of a heat transfer factor table used for calculating a room-by-room load. Table A-5 in the Appendix is an abbreviated form of Table A-4. Table A-5 is used for calculating a shell load.

Conduction.   Conduction is the process of heat transfer through a material.

Infiltration.   **Infiltration** is the term used to describe any air that leaks into a building from the outside. Because a large part of the heat loss for any building is that caused by the leakage of cold outside air into the building, the rate of the infiltration must be determined and the heat loss caused by the infiltration must be calculated. Several methods can be used to determine the rate of infiltration. One is the **crackage** method. This method requires the estimation of the **lineal footage** of the cracks around the windows and doors and the application of an appropriate factor. Another method is to figure the amount of air in the building in cu ft and assume that that much air will infiltrate the building once each hour.

Volume.   The **volume,** or cubic content, or a building is figured in cu ft (ft³) or cu m (m³) and is often used in heat

FIGURE 13–3 _____
Pictorial representation of U factor

Heat transfer factor*

Infiltration*

Crackage – The cracks around the opening parts of doors or windows.

Lineal footage – The distance between two points, measured in ft.

Volume*

Fresh air*

Ventilation air*

Air Change*

cfm*

gpm*

loss calculations. Volume is normally used when figuring the amount of air per room or zone. This variable is calculated by using the formula:

$$V = w \times l \times h$$

where:

V = volume
w = width of the space
l = length of the space
h = height of the space

EXAMPLE 13-1

If a building is 36 ft long and 24 ft wide, with a ceiling height of 8 ft, what is the volume of the building?

Solution

In this case, the dimensions are given. Therefore:
$$V = w \times l \times h$$
$$= 24 \times 36 \times 8$$
$$= 6912$$
The volume is 6912 cu ft.

Ventilation Air.   In some cases, a specified amount of outside air is brought into a building to maintain a fresh air condition for the occupants of the building. This air is called **ventilation air** or **fresh air**. Heat loss caused by this colder outside air must be figured and added to the conduction loss of the building.

Air Change.   The term **air change** relates to the quantity of air required to replace all of the air in a given space in a given length of time. One air change per hour for a building would be the amount of air required to replace all of the air in the building in one hour.

If a fresh air system is designed for a building that has a volume of 6912 ft^2, and a 1 hr air change is desired, the ventilation rate is 6912 ft$_3$ per hour. This rate of air change gives a complete change of air in the building each hour.

Shell Load.   If a heat loss for a building is figured as if the entire building is one large space, this method is called a shell load calculation.

Room-by-Room Load. If the heat loss is calculated for a building by figuring the losses for the individual rooms or separate zones and then adding these individual losses together to find the total load, this method is a room-by-room load calculation.

cfm. Air being used as the heating medium in a given system is measured in cubic feet per minute, or **cfm.** After the heat loss in Btu is calculated for a building in which air is the circulating medium, the usual practice is to convert the Btu/h figures to cfm to determine how much air to distribute to each space in the building to offset the heat loss in the space. The term cfm may be used for the air being circulated in an entire system, or it may apply only to fresh air, infiltrated air, or any other air, when the quantity must be accounted for.

gpm. When water is used as the circulating medium in a heating system, the water is measured in gallons per minute, or **gpm.** As in warm air systems, the heat loss in Btu/h is converted to gpm to facilitate the design of the piping system and the selection of heat exchangers.

13.3 HEAT LOSS VARIABLES

Designers calculating a heat load for a building must consider what variables will affect that heat loss. The two most important variables are the design temperature difference and the areas of the outside surfaces of the building.

Design Temperature Difference

The first variable, design temperature difference, occurs because buildings are constructed in different geographical locations. The outside design temperature used for calculating heat loss varies according to the geographical location of the building. A building constructed in the southern part of the United States has a smaller heat loss than does a building constructed in the same way in the North because of the difference in the outside temperature in the winter.

A set of design temperature tables is used to select the proper design temperature for the area in which a building

Dry bulb (db) temperature – The temperature measured with a typical thermometer registering sensible heat only.

is to be constructed. These tables show the winter outside design temperature according to location. An inside design temperature is chosen. The difference between the two temperatures gives the design temperature difference to use in the calculation.

Outside Design Temperature. Table A-3 in the Appendix is a complete set of outside design temperature tables for the United States. These tables show the winter **dry bulb (db) temperature** for many locations in the United States and in Canada. The figures are compiled for all cities in which there are weather stations. The designer chooses the proper temperature for the area in which the building will be located.

EXAMPLE 13-2

A designer needs to find the outside design temperature for Seattle, Washington.

Solution

The designer looks in the column under "Winter db" and across from the heading for Seattle. The temperature is found to be 17°F.

The design temperatures listed in the chart are somewhat higher than the lowest winter temperatures that occur in an area. Rather than design for the lowest temperature that may occur in an area, the listed design temperatures disregard the temperatures that occur less than 2 1/2 percent of the total winter hours. Normally the equipment chosen to heat a building has capacity enough to heat the building during colder temperatures because of heat storage in the building itself, and because the coldest temperatures occur for relatively short periods of time.

Inside Design Temperature. The inside design temperature is a matter of choice of the designer. At one time the trend was to design for 75°F or warmer inside buildings where comfort is most important. Because of the need to conserve energy and especially because of high fuel costs, the inside design temperatures now may be figured at 65°F to 70°F, depending upon the building use.

The design temperature for manufacturing plants and buildings used for storage, such as warehouses, may be much lower than that required in a residence. The choice of temperature for these buildings should be made with the uses of the building in mind.

Calculating Design Temperature Difference

The design temperature difference (DTD) to be used in the heat loss calculation is found by subtracting the outside design temperature from the inside design temperature.

EXAMPLE 13-3

The winter db for Seattle, Washington, is found in the design table to be 17°F. If the inside design temperature of 70°F is chosen, what is the design temperature difference?

Solution

Subtract 17°F from 70F.
    70 − 17 = 53

The design temperature difference to be used for figuring a load for the Seattle area is 53°F.

Exposed Surfaces

The second variable, the areas of the exposed surfaces of the building, occurs because no two buildings are ever exactly alike. The outside surfaces of windows, walls, ceilings, and other components will be different for each building. The areas of exposed surfaces for the different parts of the building through which heat can pass have to be considered. The dimensions of the surfaces for each different material combination that is exposed to outside temperatures must be determined, and then the area of each calculated.

Calculating the Area of Exposed Surfaces

The area of a surface is figured in sq ft (ft^2) or sq m (m^2). The area of a surface is calculated by using the formula:

$$A = w \times l$$

where:

A = area

w = width of surface

l = length of surface

EXAMPLE 13-4

If a building is 36 ft long and has a ceiling height of 8 ft, what is the gross wall area of one side?

Solution

In this case, the ceiling height of the building is used as the width in the formula. Therefore:

$$\begin{aligned} A &= w \times l \\ &= 8 \times 36 \\ &= 288 \end{aligned}$$

The gross wall area is 288 sq ft.

The most common building components for which the areas have to be figured are

1. Exposed walls
2. Windows
3. Doors
4. Ceilings
5. Floors
6. Cold partitions

In the case of an existing building, the dimensions may have to be taken from the building itself by measurements. The best way to keep a record of the dimensions is to draw sketches of the building and record the dimensions. A floor plan should be sketched showing the exterior walls and the interior partitions and doors. Elevation sketches should be made showing the exterior doors, windows, and roof details. The dimensions of the various areas to be considered in the heat loss are shown on the sketches. Construction types and details can also be recorded on the sketches to help in selecting the proper factors for the loads. The sketches become a permanent part of the job file and can be used in the system design process, as well as for figuring the loads. In the case of a new building, plans usually are available giving dimensions.

EXAMPLE 13-5

A typical double-hung window size is 2'4" by 5'4". Find the area.

Solution

The dimensions of doors and windows can be figured to the nearest half-foot for purposes of finding the area. Therefore, the dimension 2'4" by 5'4" rounded off to the nearest half-foot is 2'6" by 5'6". Changing the figures to decimal equivalents makes the math simpler. In this case 2'6" is written as 2.5 and 5'6" as 5.5. To calculate the area of the window, use the formula:

$$
\begin{aligned}
a &= w \times l \\
&= 2.5 \times 5.5 \\
&= 13.75
\end{aligned}
$$

The area of the window is 13.75 sq ft. Working with the decimal equivalents makes the calculations of the window areas simpler, and the results are still accurate enough for purposes of determining the heat loss.

EXAMPLE 13-6

Find the area of a wall that is 24'9" long. The ceiling height is 8'3".

Solution

The dimensions of walls, ceilings, and floors can be figured to the nearest whole foot for purposes of finding the area. Therefore, the dimensions 24'9" and 8'3" rounded off to the nearest whole foot become 25' and 8', respectively. To calculate the area of the wall use the formula:

$$
\begin{aligned}
A &= w \times l \\
&= 25 \times 8 \\
&= 200
\end{aligned}
$$

the area of the wall is 200 sq ft.

FIGURE 13–4
Duct system around the perimeter of a building under the cement slab floor

Heat loss is figured on the net exposed wall only. However, in figuring the areas, it is easier to calculate the area of the entire wall, or gross wall, and then subtract the window and door areas from **gross wall area** to find the **net wall area.**

If the exposed walls of the building are made up of two or more kinds of construction, then the areas of each kind of construction must be figured separately since each kind of construction is likely to have different heat loss factors.

Closets, halls, and stairways should be treated as if they are part of the rooms into which they open.

A **cold partition** is defined as a partition that separates a heated space from a space that is not heated but is enclosed and is not exposed to the outside temperature. Such a space generally is not as cold as the outside but not as warm as the inside. An example of a cold partition is the wall between a house and an enclosed porch or vestibule.

A **cold ceiling** is the ceiling of any room that is located beneath an unheated space. This space can be located beneath an unheated space. This space can be an attic, the outdoors, or any other space exposed to outside temperatures.

A **cold floor** is any floor between the heated part of the house and an unheated space, such as a crawl space, a cold garage, a space under an overhang, or any other such space exposed to the outside temperature.

Concrete floors on the ground can pose a special problem for figuring heat losses. If the floor is in a building that is heated by an overhead system, then the floor should be treated as though it were a separation between the room and a cold area (in this case, the ground), so the area of the floor should be figured. It is necessary only to figure the outside 10 ft of the floor, however, since the heat loss through the floor farther than 10 ft from the wall is negligible.

If the concrete floor has heat ducts around the **perimeter** in the slab itself, some heat will be lost from the duct. This lost heat will warm the outside edge of the floor. See Figure 13-4 for an illustration of a duct system under a concrete slab around the perimeter of a building. In this case, the only measurement required for the floor loss is the perimeter length of the slab, which is adjacent to the outside wall.

Perimeter – Circumference or distance around.

13.4 HEAT LOSS FACTORS

The factors to consider when determining heat loss are the rate of heat conduction through the exposed surfaces of the building and a rate of cold air infiltration. Selecting the proper factors for each of the materials used in building construction is one of the most important steps in the calculation of heat loss. The factors are the numerical equivalents of the Btu/h that will be conducted through the building components, per sq ft of the area, per 1°F difference in temperature on the two sides of the material. There are also factors for the heat loss due to the air that infiltrates into a building through the cracks around the opening parts of the windows and doors and that gets into the building when the doors are opened.

Factor Tables

A reasonably complete set of factors, or factor tables, should include factors for not only the different components in the building, such as walls, ceilings, and floors, but also for different types of construction for each. The set also should include factors for different amounts of insulations as used in the major construction types. The factor tables found in this text are heat transfer factor tables. In heat transfer factor tables, the U factors for various types of materials have been multiplied by the design temperature difference. U factor tables also are available. A complete set of U factor tables is available in the current manual of the American Society of Heating, Refrigeration, and Air Conditioning Engineers, Inc.

Most factor tables also have a selection of factors to be applied to door and window crackage to determine the loss due to infiltration of cold air. Some tables, however, include the infiltration factor in the conduction factor for the windows, so designers must check the particular tables being used and make sure the infiltration is included in one place or another.

Conduction

Selecting the factor for conduction for any part of the building requires finding the description of the component in the tables that most nearly matches the construction. Under the heading for exterior walls in the factor tables are found major headings for masonry walls for various types, frame walls of different types, and possibly others.

Under each major heading, subheadings for variations are found. These variations are thickness, finish material, amount of insulation, and any other factor that affects the heat flow through the building component. All of these variations must be taken into consideration when selecting the factor so that the final loss figure will be correct for that particular construction.

The selection of the proper factors for windows, doors, walls, ceilings, and floors is made in the same way.

Infiltration

When selecting the factor for infiltration, the type of window or door for which the loss is being figured must be identified.

No matter how tightly a building is constructed, air will leak through cracks around windows and doors, as well as through cracks that may exist between building components. Air at outside temperatures will leak into the building through these cracks and has to be heated if the building is to be maintained at a fixed inside temperature. The air leakage is caused by the wind pressure against the exterior surfaces of the building. The amount of leakage is determined by the size of the cracks and the velocity of the wind. In the calculations, the size of the cracks is compensated for by the factors used for infiltration, and the wind velocity is figured as 15 miles per hour. (Incidentally, a fireplace can increase the infiltration into a building appreciably because the fireplace requires combustion air to burn, and that air usually comes from crackage in the construction.)

Calculation of the heat loss for infiltration uses the actual footage of the cracks in each door or window that has an opening to the outside. The factor tables give the proper factors to use with these lineal footage figures to arrive at the heat loss occasioned by the cracks.

In measuring crackage, the lineal footage is figured for only the part of the window or door that opens. An illustration of the opening crackage for a double-hung window is shown in Figure 13-5. The cracks around the stationary,

FIGURE 13-5
Opening crackage for a double-hung window

or nonopening, parts of the windows or doors do not have to be figured.

Ventilation Air.   In some cases, outside air air is introduced to a building as ventilation air, **or make-up air** for an exhaust fan or a fireplace. This fresh air is introduced into the return air system, as shown in Figure 13-6.

When fresh air is introduced, the heat loss for this air has to be figured in with the heat loss caused by conduction losses. To calculate the loss, the quantity of air that is introduced is calculated in cfm. Then the loss is figured on that air by use of the sensible heat formula, as follows:

$$\text{Btu/h} = \text{cfm} \times 1.08 \times TD$$

when:

Btu/h	=	amount of heat required to raise or lower the temperature of the air
cfm	=	quantity of air in cu ft per min
1.08	=	a factor including adjustments for the specific heat of air, the specific weight, and time
TD	=	temperature difference

EXAMPLE 13-7

If 200 cfm of outside air is brought into a building as ventilation air and the temperature difference is 53°F, find the heat loss.

Solution

$$\begin{aligned}
\text{Btu/h} &= \text{cfm} \times 1.08 \times TD \\
&= 200 \times 1.08 \times 53 \\
&= 11{,}448
\end{aligned}$$

The heat loss caused by the fresh air is 11,448 Btu/h.

13.5 SUMMARY

Designers must make an accurate calculation of the heat load of a building so that they can select the proper size heating unit and design the most efficient heat distribution system. To calculate a heat load on a building, all of the variables including temperature difference between the

Make-up air – Air that is brought into a building specifically to make up any air that is being vented.

FIGURE 13–6
Introduction of fresh air into the return air system

inside and outside, and the areas of surfaces exposed to outside air. Heat loss factors include the rate of heat conduction through the surface of the building and the rate of cold air infiltration.

Traditionally, heat loads were figured manually by combining the variables and factors for each building seperately, and mathematically figuring the losses for each part of the building. Now, most heating and cooling loads for buildings are calculated using computers and special load calculation programs.

Designers should follow certain steps when calculating a load.

- Determine design temperature difference.
- Calculate the areas of each type of construction material used in the building.
- Calculate infiltration or fresh air in cfm.
- Determine the proper factors.
- Calculate the heat loss for each component part of the building.
- Add the component losses together for the total heat load.

The design temperatures are determined by reference to an outside design temperature table and by arbitrary selection of the inside temperature. The design temperature difference is found by subtracting the outside design temperature from the inside design temperature.

The areas of each of the building components, such as windows, doors, ceilings, floors, and walls, are calculated from measurements made of the actual building or are taken from plans of the building.

Selection of the proper factors for conduction and infiltration is made from an appropriate factor table that covers the type of construction used in the building. Consideration is given to building materials, methods of application, and insulation in selecting the factors. The final result, the total heat load, is the sum of the heat losses through each of the component parts of the building and the heat required to raise the infiltrated, or fresh, air to the temperature of the indoor air.

13.6 TERMS TO KNOW

heat load
heat loss
heat gain
variable
factors
components
temperature
design temperature
outside design temperature
inside design temperature
design temperature difference
R factor
U factor
conduction
heat transfer factor
infiltration
crackage
lineal footage
volume
fresh air
ventilation air
air change
cfm
gpm
dry bulb (db) temperature
shell load
room-by-room load
gross wall area
net wall area
cold partition
cold ceiling
cold floor
perimeter
make-up air

CHAPTER
14
Calculating Heat Loads

Heat load calculation form*

Shell load*

Room-by-room load*

*Terms defined in text.

14.1 INTRODUCTION

Organizing the heat loss variables and factors in a format that makes it easy to use them for calculations and then combining them in a way to make those calculations is called figuring the heat load for a building.

The forms used for calculating heat load are called **heat load calculation forms.** In this chapter, two types of forms that can be used for heat load calculations are described for calculating loads manually, and one form for calculating loads by computer. The forms for manual loads: short form for calculating shell load and the long form for calculating room-by-room load.

14.2 CONVENTIONAL LOAD CALCULATION FORMS

The first step in manually calculating a heating load is to identify the heat loss variables and factors that will be used for the calculations as described in Chapter 13. The heat load calculation form is used to record the data in an organized way. The form with the data recorded is then used to make the necessary calculations.

If the load is figured for one space in a building, or for any number of spaces treated as one, it is called a **shell load;** and a short form calculation sheet for figuring a shell load is used. This form has one column for the areas, another column for factors, and a third column for the heat loss of the separate components of the space. The component losses in the third column are added together to get the total building loss.

If the load is being figured for many rooms or many spaces, with each being treated as a separate part of the whole, it is called a **room-by-room load,** and a long form calculation sheet for figuring the room-by-room load is used. The long form has one column for the factors that remain the same for all components in the building and separate columns for each room and the variables related to the rooms, such as the size of windows, doors, walls, ceilings, and floors. This form allows a determination of separate loads for each room. These individual loads are added together to get the total building loss.

Any calculation form used should include space at the top to record the job name, address, date, and, most important, the design temperatures used for the load. Since the heating loads are used for sizing the equipment and designing the distribution system, the forms are valuable documents and should be kept on file for reference.

Short Form

A typical short form calculation sheet for figuring shell load is shown in Figure 14-1 and also in Table A-6 in the Appendix. The spaces in the top section of the form are used for the statistical data for the particular job. The middle section of the form is divided into horizontal and vertical divisions for easy calculations of the loads. The bottom of the form provides space for total loads. The first column on the left identifies the various components of the building. The first several lines refer to building components. The last line is for the loss due to air that infiltrates the building or that is introduced as fresh air.

The second column, area, is for the areas of the spaces identified in the first column or for infiltration in cfm, or both.

The third column, htf, is for the heat transfer factor as taken from the factor tables. If U factors are used, the U factor must be multiplied by the design temperature difference in order to be used as a multiplier, or heat transfer factor, on this form.

The fourth column is for the subtotal Btu/h figures. Btu/h is arrived at by multiplying the areas, or cfm, times the htf. The subtotals are then added together to arrive at the total Btu/h loss, shown on the last line.

Long Form

Whenever the heating load calculation is used for selecting the supply and return registers and grilles and, for sizing the duct system, the loads should be figured on a room-by-room basis, or at least on a zone-by-zone basis, so the supply air can be distributed in the proper quantity to each space, to provide for the heating of the space.

To facilitate the calculations of loads on a room-by-room basis, a long form calculation sheet is used.

Portions of the long form are shown in Figure 14-2 and the complete form is shown in Table A-7 in the Appendix.

The informational heading on the long form are similar to those used on the short form. The first column on the left-hand side of the page, in which building components

JOB NAME _____ DATE _____

ADDRESS _____   } Step 1

DEALER _____

O.S. Design Temp _____ I.S. Design Temp _____
Design Temp Diff _____
Wall Construction _____ Insulation _____ " Window Type _____   } Step 2
Roof Construction _____ Insulation _____ " Floor _____

	AREA	HTF	BTUH LOSS		
GROSS WALL		✕	✕		Step 3
WINDOWS					Step 4
DOORS					
NET WALL					Step 5
CEILING					Step 6
FLOOR					Step 7
CRACKAGE					
INFILTRATION W / D					Step 8
TOTAL LOSS	✕	✕			Step 11

Step 9   Step 10

FIGURE 14-1
Short form calculation sheet for figuring shell loads

are identified, also is similar. The difference in the two forms is that to the right of the column for the multiplier, the long form is divided into multiple columns, one column for each room. A subheading across the top and separate columns for each room make it possible to figure the loss for individual rooms or spaces.

DATE .... BY ....

OUTSIDE DESIGN TEMP. ....

HEATING CONTRACTOR .... INSIDE DESIGN TEMP. ....

NAME OF JOB .... DESIGN TEMP. DIFF. ....

	Table No.	"U" Factor	Multi-plier	Area Crack Cfm	ʰuh ss	Area Crack Cfm	Btuh Loss	Area Crack Cfm	Btuh Loss	Area Crack Cfm	Btuh Loss	Area Crack Cfm	Btuh Loss
1. Room Use and Floor Level	1.			6.		7.		8.		9.			
2. Room Length and Width													
3. Running Ft of Exposed Wall													
4. Ceiling Height													
5. Exposure Type													
Gross 6. Exposed Wall													
Windows 7. and Doors													
Net 8. Exposed Wall													
9. Cold Partition													
10. Cold Ceiling													
11. Cold Floor													
Infiltration, 12. (Leakage) or Air Change													
13. Room Sub-Total Btuh													
14. Btuh Adjustment													
15. Room Total Btuh													

FIGURE 14-2

Portions of long form calculation sheet used for figuring room-by-room loads

Calculation Procedures for Finding Heat Load

The mechanics of calculating a heating load are simple. When using the short form, the designer lists the areas (A) of the building's components through which losses occur, selects the proper heat transfer factor (htf) for the design temperature difference, and then multiplies one by the other to obtain the total heat loss (H) in Btu/h. The answer gives the loss for each of the different components of the building. The component losses are then added, together to get the total loss.

The formula applied for each of the components individually is as follows:

H $= A \times htf$

where:

H   = Btu/h
A   = area of the component
htf   = heat transfer factor from the factor table

Remember: For this calculation, the htf is selected for a particular design temperature difference and includes the proper U factor for the material.

EXAMPLE 14-1

The second line on the short form is used for figuring the heat loss through the windows. Calculate the heat loss through the windows for a house with 204 sq ft of window area. Use an htf of 68.

Solution

$$H = A \times hft$$
$$= 204 \times 68$$
$$= 13{,}872$$

The heat loss through the windows is 13,872 Btu/h.

In the long form for calculating a heat load, the short form formula can be used or the U factors can be used. A U factor relates to the heat loss for only 1°F of temperature, and so the design temperature difference also must be included to achieve the heat transfer factor. If the U factor is used on the long form, the following formula is used for each separate component:

H $= U \times DTD \times A$

where:

H   = total heat loss in btu/h
U   = U factor for the material
DTD   = design temperature difference
A   = area of the component

EXAMPLE 14-2

On the long form calculation sheet, in Section 7, for windows and doors, the second column from the left has a place for the U factor and the third column has a place for a multiplier. The multiplier in this case is the U factor times whatever design temperature difference applies to this job.

A window in a building is 8 ft by 5 ft in dimensions, which makes it 40 sq ft in area. The window has no storm sash. Referring to Table A-5 of the Appendix, the U factor is 1.13. If, on a particular job, the design temperature difference is 60°F, what is the heat loss through the window?

Solution

H $= U \times DTD \times A$
     $= 1.13 \times 60 \times 40$
     $= 2712$

The heat loss through the window is 2712 Btu/h.

14.3 CALCULATING A SHELL LOAD

In a shell load calculation, the building loss is figured as though the building were one large room. Since the short form calculation sheet is used when making calculations for one space, it is ideal for figuring shell loads. The following section describes the steps for calculating the shell load for a building using the short form calculation sheet. Each step is identified with a label on the short form shown in Figure 14-1.

Statistical data – Identifying data of record, such as job name and address

Perimeter distance*

Job Data

Step 1 in calculating heat load, with any form, is to fill in the proper **statistical data** for the job on the top of the form.

Construction Data

Step 2 in calculating heat load is to fill in the proper data relative to the construction of the building and the design temperatures that will be used. It is important that a record be kept of the data used in the selection of factors in case any question arises concerning the loads.

Gross Wall Area

Step 3 in calculating heat load on the short form is to find the gross wall area. The gross wall area is found by multiplying the distance around the building, called the **perimeter distance,** by the wall height. The resulting figure is entered in the area column, across from the gross wall designation on the form.

Window and Door Areas

Step 4 in calculating heat load on the short form is to determine the window and door areas. To figure the areas of windows and doors, multiply the width by the height. If the dimensions are rounded off to the nearest half-foot and the totals rounded off to the nearest whole foot, the accuracy will be sufficient for the calculations. The resulting figures are entered for each in the area column, across from the proper designation.

Net Wall Area

Step 5 in calculating a heat load on a short form is to determine the net wall area. The combined area of the windows and doors is subtracted from the gross wall area to arrive at the net wall area. The net wall area is entered in the area column, across from the net wall designation.

Ceiling Area

Step 6 in calculating a heat load on the short form is to determine the ceiling area. The area of the ceiling is figured by taking the length of the building times the width of the building. This figure is entered on the form, across from the ceiling designation.

Floor Area

Step 7 in calculating a heat load on the short form is to determine the floor area. The floor area usually is the same square footage as the ceiling square footage. An exception occurs when a floor overhangs beyond the normal outside wall line, or when a cathedral ceiling is used. A cathedral ceiling is one in which the ceiling material is directly attached to the building rafters. In these special cases, the floor and the ceiling areas must be figured separately. The floor area can be entered in the area column, across from the floor designation.

Infiltration or Fresh Air Loss

Step 8 in calculating heat load on the short form is to determine the infiltration losses. The most accurate way to figure the loss for infiltration air is to estimate the length of the crackage, in lineal feet, that occurs around the opening sections of the windows and doors. The length of the crack for each window, and door is entered in the column under "crackage," across from either the window (W) or door (D) designation.

The length of the crack will vary according to the size and type of window. Double-hung windows have two openers to each window. The top sash can be opened downward, and the bottom sash can be opened upward. The crackage is figured for each sash separately. The two figures are added together to get the total crackage for each window.

If the windows are horizontal sliders, usually only one side of the window is an opener. The crackage occurs around the opening part of the window only. See Figure 14-3. Whatever the style of window used, the designer should figure the crackage on the opening cracks only.

FIGURE 14–3
Horizontal slider window showing the stationary and opening parts

Crackage around doors is figured in the same way as it is for windows. There are not as many kinds of doors as there are windows. If the crackage is being figured for double doors, it is necessary to include both doors only if both open.

In some cases, ventilation air, or make-up air, will be introduced into a building. When such is the case, the heat loss for the air introduced is figured instead of the loss for infiltration. The quantity of air is calculated in cfm, and the loss is figured by the sensible heat formula. This loss rather than infiltration loss is used on the form.

Heat Transfer Factors

Step 9 in calculating heat load on the short form is to find the proper heat transfer factor from a set of htf tables for each of the building components. A typical factor table is shown in Table A-5 in the Appendix. The factor selected for each component is entered on the heat loss form, in the htf column, across from the appropriate building component.

The heat transfer factor table is an abbreviated table of factors for the most common types of building components used on residences in the western United States. Although the table is by no means complete, it can be used for any building component described in the left-hand column. This column describes some of the common combinations of materials normally used as components in construction. Windows, doors, and walls are listed here. The list also includes some choices for different construction methods that may be employed.

The other vertical columns on the table show the proper heat transfer factors to be used for the different design temperature differences. The column at the extreme right shows the U factor for the type of construction shown on the left.

EXAMPLE 14-3

A window with no storm sash is being used. The design temperature difference is 53°F, as found for Seattle, Washington, at a 70°F inside design temperature. Find the htf.

Solution

Using Table A-5 in the Appendix, the htf is found in the

50°F column (because this is the closest figure to 53°) and across from "glass, no storm sash." The heat transfer factor is 57. The figure 57 should be entered in the htf column for the window on the short form.

The same selection process is used for each of the other components of the construction, such as doors and walls, and the proper heat transfer factors are then entered on the form in the htf column.

The two sections at the bottom of Table A-5 in the Appendix contain factors for heat loss due to infiltration of outside air through the cracks around the openings of windows and doors. These factors give the Btu/h loss per foot of crack for the design temperature differences found at the heat of the columns.

The factors for the windows are for average-fit, weather-stripped windows, and are appropriate for most of the good-quality windows in use today. The door factors are separated into two categories, for two different types of door installations. The first is for an average-fit, nonweather-stripped door. The second is for a weather-stripped door, or a door and storm door combination. The proper factors should be selected according to the type of construction used and then listed on the loss form in the htf column.

Subtotals and Total Btu/h

Step 10 in calculating the heat load on the short form is to determine the Btu/h subtotal for each component. After the areas of all of the components of the construction are figured and entered in the second column of the short form calculation sheet and the multiplying factors are selected and entered in the third column, it is necessary only to multiply the areas times the factors to arrive at the Btu/h subtotals for each component of the construction. The subtotals are then recorded in the third column on the form.

Step 11 is to add the subtotals to find the total Btu/h loss of the building. The total Btu/h loss is also the total heat load of the building.

EXAMPLE 14-4

Use the construction details on the floor plan shown in Figure 14-4 and the short form calculation sheet to calculate the heat load for the residence. An illustration of a short form filled out with the data for the example is shown in Figure 14-5.

Steps 1 and 2:   Record the job data and construction data at the top of the form. To determine the DTD, figure the load for Salt Lake City, Utah. The outside design temperature for Salt Lake City is -2°F, as found on the outside design temperature chart. Use an inside design temperature of 70°F. The design temperature difference then is 72°F:

$$DTD = 70°F - (-2°F) = 72°F$$

FIGURE 14–4

Sketch of floor plan of a typical house with necessary data recorded for figuring a heat loss

Step 3: Figure the gross wall area by multiplying the perimeter distance around the house by the ceiling height:

perimeter distance = $(2 \times 48) + (2 \times 30) = 156$
ceiling height = 8
gross wall area = $156 \times 8 = 1248$

JOB NAME _____ DATE _____

ADDRESS _____

DEALER _____

O.S. Design Temp _____-2° F_____ I.S. Design Temp ___70° F___
Design Temp Diff ___72° F___
Wall Construction ___FRAME___ Insulation __2"__ " Window Type _____
Roof Construction ___W/C'LG.___ Insulation __4"__ " Floor ___CRAWL___

	AREA	HTF	BTUH LOSS	
GROSS WALL	1248			
WINDOWS	158	79	12,482	
DOORS	42	22	924	
NET WALL	1048	10	10,480	
CEILING	1440	7	10,080	
FLOOR	1440	10	14,400	
	CRACKAGE			
INFILTRATION W	47 lin ft	42	1974	
D	21 lin ft	70	1470	
TOTAL LOSS			51,810	

FIGURE 14–5 _____
Completed short form calculation sheet for the house in Figure 14-4

FIGURE 14–6
Step-by-step procedure for figuring a long form calculation sheet

The gross wall area is 1248 sq ft. Place this figure in the gross wall area space on the form.

Step 4: Figure the window area by adding up the separate areas of all the windows in the house (figure the sliding glass door as a window):

window area = 158

Place 158 sq ft in the window area space.

Find the total door area by adding all wooden door areas (round off to the nearest whole foot):

door area = 2 × 21 = 42

Place 42 sq ft in the door area by subtracting the door and

HEATING DATA SHEET

JOB NO.
DATE BY
OUTSIDE DESIGN TEMP. −2° F
INSIDE DESIGN TEMP. 70° F.
DESIGN TEMP. DIFF. 72° F

HEATING CONTRACTOR ADDRESS

NAME OF JOB ADDRESS

	Table No.	"U" Factor	Multiplier	1. L.R Area/Crack/Cfm	Btuh Loss	2. BR #1 Area/Crack/Cfm	Btuh Loss	3. BR #2 Area/Crack/Cfm	Btuh Loss	4. BR #3 Area/Crack/Cfm	Btuh Loss	5. BATH+½B Area/Crack/Cfm	Btuh Loss	6. UTILITY Area/Crack/Cfm	Btuh Loss	7. KITCHEN Area/Crack/Cfm	Btuh Loss	8. D.R. Area/Crack/Cfm	Btuh Loss	9. HALL Area/Crack/Cfm	Btuh Loss
1. Room Use and Floor Level				1. L.R		2. BR #1		3. BR #2		4. BR #3		5. BATH+½B		6. UTILITY		7. KITCHEN		8. D.R.		9. HALL	
2. Room Length and Width				18 x 16		12 x 13		14 x 13		14 x 14		8 x 14		8 x 14		8 x 14		10 x 14		20 x 3	
3. Running Ft of Exposed Wall				34		12		27		28		8		8		8		24		0	
4. Ceiling Height				8		8		8		8		8		8		8		8		8	
5. Exposure Type																					
6. Gross Exposed Wall				272		96		216		224		64		64		64		192		0	
7. Windows and Doors — W.		79		40	3,160	18	1422	18	1422	18	1422	12	948	0	~	16	1264	36	2844	0	
D.		22		21	462									21	462						
8. Net Exposed Wall		10		211	2110	78	780	198	1980	206	2060	52	520	43	430	48	480	156	1560	0	
9. Cold Partition																					
10. Cold Ceiling		7		288	2016	156	1092	182	1274	196	1372	112	784	112	784	112	784	140	980	60	420
11. Cold Floor		10		288	2880	156	1560	182	1820	196	1960	112	1120	112	1120	112	1120	140	1400	60	
12. Infiltration (Leakage)		42		18	756	12	504	12	504	12	504	10	420	0		12	504	18	756	0	
or Air Change		70		21	1470									21	1470					0	
13. Room Sub-Total Btuh					12 854		5358		7,000		7318		3792		4266		4152		7540		420
14. Btuh Adjustment																					
15. Room Total Btuh																					
16. Cfm @ Temp. Rise																					
17. Adjusted cfm																					

FIGURE 14–7

Completed long form calculation sheet for the house in Figure 14-4 window areas from the gross wall area:

net wall area = 1248 - (158 + 42) = 1048

Place 1048 sq ft in the space for the net wall.

Step 6 and 7: Find the ceiling and floor area. To find this figure, multiply the length of the house by the width:

ceiling and floor area = 48 × 30 = 1440

Enter 1440 sq ft in the spaces for both the ceiling and the floor.

Step 8: Find the crackage on doors and windows to determine infiltration. The length of the crackage for the windows needs to be figured for this calculation. Since the windows are horizontal sliders, one-half of each can be expected to open. Figure the lineal feet of crack for each window and for the sliding door. Add the window and sliding door totals together. This is the total window crackage. Because this example is for a shell load, only half of the crackage is used. Take one-half of the crackage and place the number under "crackage," across from "windows" on the form. Do the same calculation for the wood doors:

window infiltration crackage = 94 ÷ 2 = 47
door infiltration crackage = 42 ÷ 2 = 21

Enter 47 lineal ft for the window crackage and 21 lineal ft for the door infiltration.

Step 9: Find the appropriate htf for each of the surfaces and the infiltration from the factor table in Table A-5 in the Appendix. Enter the data in the htf column on the form.

Note: Although the design temperature difference is actually 72°F, the 70°F column on the factor table will give accurate enough results.

Steps 10 and 11: Multiply the areas in Column 2 by the factors in Column 3. The results are the component losses. Add the component losses together to arrive at the total heat loss. The total is 51,810 Btu/h, as seen in Figure 14-5.

14.4 CALCULATING A ROOM-BY-ROOM LOAD

The long form calculation sheet shown in Figure 14-2 is used when the designer wants to figure a load on a room-by-room basis. The procedure is similar to that used for figuring a series of shell loads. The final results are a load for each room in the building, but the format of the long form makes figuring the loads relatively easy.

14.5 CALCULATING A LOAD BY COMPUTER

Computer load calculations normally include both heating and cooling loads for a building. The computer program for figuring the loads is contained on a computer diskette. The program is written so it contains permanent data and related factors for load calculations. To run a heating load with a computer, the diskette that has the program on it is inserted in a disk drive in the central processing unit (cpu) of a computer, where the program is taken from the diskette, transferred to the central processing unit, and is ready for use. During operation, the program asks for information concerning any building for which loads are required. An operator inputs variables concerning the building. The computer system performs the mathematical calculations for the load, and produces output concerning the load.

A computer system for figuring load calculations consists of various computer devices. These make up what is called the hardware of the system.

Hardware

Hardware consists of all physical units of a computer system. A mininal system includes a computer (cpu), keyboard, disk drives, monitor, and printer.

A **computer,** or cpu, is the control center of the computer system. It contains the electronic components and circuitry that processes information that is received from the input devices, and outputs results to the output devices.

A **keyboard** is an input device by which commands can be input to the computer system. The input data is translated by the computer into electronic signals that are processed within the cpu.

Disk drives are mechanical devices that function similarly to a turntable in a record player. The diskette that has the load calculation program written on it is inserted in a disk drive. A magnetic reader takes the program off the disk, and it is transferred to the computer processing unit.

A computer **monitor** is a video screen device that shows data that has been input as well as output data relative to a program being run. When a load calculation program is run, input data includes variables and some factors for the job run. Output data are the heating and cooling loads for the job.

A computer **printer** is a printing device similar to a typewriter. Output attained by running a program is printed on a computer printer, on paper, to provide a more or less permanent copy of output data.

Software

Software is the program that runs a computer system. The software directs the operation of the computer, while running a specific program for a specific purpose. A heating and cooling load calculating program contains the data for calculating heating and cooling loads on a particular building, and providing output concerning those loads.

A load calculation program requires input of some variables and factors related to a specific building for which loads are desired. The speed of operation of a computer greatly reduces the time required for calculations, and generally will furnish more data concerning heating and cooling loads than can easily be acquired by running the loads manually. An operator running heating and cooling loads on a computer should be familiar with the conventional load calculation process.

Heating and cooling load calculation programs are available from engineering and technical associations, manufacturers of heating and air conditioning equipment, and private computer programming companies. Instruction manuals come with the various programs. The manuals describe the program, and also give step-by-step instructions for running the program.

Most load calculating programs include a job information sheet which is filled out with the variables related to a job before the program is run. Data from the job information sheet is input to the computer for figuring loads.

Running a Program

To run a heating and cooling load calculation program, a software diskette containing the program is inserted in a disk drive. The disk drive loads the program from the disk into the memory of the cpu. As the program runs, the computer asks for information relative to the job for which the loads are being run. The input information is taken from the job information sheet that has been filled out ahead of time.

In most load calculation programs, questions that ask for input appear on the monitor. The input is data required for the calculations. The questions usually begin by requesting general information such as the job name, geographical location, and design temperatures. The next series of questions request the sizes of the building, rooms, or zones. Some programs have factors related to different types of construction stored on the disk. Questions relating to factors are about construction methods and materials. Programs that do not have factors on the disk ask for them from the operator. The factors are the same as those found on tables of factors used for manual load calculations.

As a load calculation program runs, the operator inputs response to questions that appear on the monitor. Explanations about procedure appear on the monitor, and many programs have options that give help when requested. A program instruction manual should answer any question that an operator has about a program.

After loads have been calculated, the results are shown on the computer monitor. The monitor display includes a breakdown that shows subtotals for the component loads, and for individual rooms or zones. Some programs also display information relating to the job variables, such as areas by exposure, infiltration data, and people and equipment input data. This information is used for checking the accuracy of the loads. A printout of the loads is made at this point on the computer printer. The printout may be used as a worksheet for checking the loads, and also is a permanent copy of the load calculation.

14.6 ADJUSTMENTS TO THE HEAT LOAD

After calculating the total building load, the designer should make two adjustments before using the figure to do any sizing. If there are any losses that have not been included in the original calculations, they should be added to the total. If there are any internal **heat gains** from light, people, or equipment or external gains from solar sources, these gains may be subtracted from the total.

Other Heat Losses

On a residential job, some losses that may not have been included on the original calculations are those caused by a fireplace and by the exhaust fans in the bathrooms and the kitchen. Most fireplace dampers are leaky even when closed. The leakage must be made up by increased infiltration to the house. The loss normally can be compensated for by adding 1000 Btu/h loss to the total load for each fireplace in the house. The loss caused by the exhaust fans is of the same type, that is, the air exhausted by the fans has to be made up by infiltration, which causes extra heat loss. To figure the Btu/h for this loss, take the cfm capacity of the fan and multiply the cfm by the hourly average use. Figure the load on that much infiltration and add it to the building total. The capacity of the fan in cfm can be found by reference to the manufacturer's literature.

The main cause of extra heat loss for commercial or industrial buildings usually is fresh air or make-up air brought in by the ventilation system. This loss should have been included in the calculations originally.

Heat Gains

Heat gain to a building during the heating season can be substantial. Any heat gain that is constant can be deducted from the loss to find the net loss to be covered with the heating unit.

In a residence, heat gains such as **solar gain,** heat from the people in the building, or heat from the lights is considered a bonus as far as heating the building is concerned. Since it is difficult to predict accurately when these gains may occur, a designer usually would not deduct them from the total heating load.

The internal heat gain can be calculated much more accurately in buildings used for commercial or industrial purposes because the gains are more constant. Normally, a predictable number of people will be in the building, and the gain from lights and equipment is much easier to figure than in a residence. All such gains can be calculated and deducted from the loss. Variable gains, such as solar gain, cannot be predicted with enough accuracy by the methods covered in this text to deduct from the loads. On large jobs, gains often are figured by using a computer, and even the variable gains can be estimated with enough accuracy to be deducted with confidence.

Duct Heat Loss

At this stage of the design, a preliminary layout of the duct system should be made. If any of the ducts are to run through any space that will not be heated, the heat loss through the walls of the ducts must be added to the total heat loss for the building. If the ducts run within the heated space in the building, any loss through the duct walls will help heat the building, and no additional loss is experienced.

Depending on where the ducts run in the building, the loss through them may vary from 0 percent to 40 percent of the total loss of the building. The designer must select the proper percentage figure to use and adjust the total loss to compensate.

Three factors must be taken into consideration in selecting the percentage to use:

1. What is the design heat loss of the building in Btu/h per sq ft?
2. Is the duct in a ventilated area but exposed to outside air (such as an attic) or is it in an unheated area (such as an enclosed basement)?
3. Is the duct insulated?

Reference to a duct heat loss allowance table, such as the one shown in Table 14-1, will make the selection easy.

On the left side of the table, the two possible duct locations are listed. The designer selects the location that describes the duct placement for the particular job being

TABLE 14–1
Duct Heat Loss Allowance Table

| | Building heat loss more than 30 Btu per sq ft | | Building heat loss less than 30 Btu per sq ft | |
	Temperature below 10°F	Temperature above 10°F	Temperature below 10°F	Temperature above 10°F
Duct Locations				
Ventilated area (attic or crawl space)	40%	30%	20%	10%
Unheated space (basement)	25%	20%	10%	5%

considered. Then the designer selects the category under building heat loss and outside designer temperature that best fits the job. The figure that corresponds to the designer's selection indicates the percentage of the total load that is lost through the ducts for that portion of the building. This percentage figure is used to find the Btu/h that should be added to determine the correct heat loss figure.

This method of correcting the total loss figure to include duct heat loss should be used only when the duct is inside the building shell. If the duct work runs outside the building and is exposed directly to the outside air temperature, the heat loss through the walls of the duct should be figured the same as the loss through any other building component.

14.7 ACCURACY

A heat loss calculation for any building is only as accurate as the data used to figure it. The data for the calculations should be selected with great care and the actual calculations made carefully. If the designer thinks that adjustments should be made so that the calculations are correct, such adjustments should be made to the factors and not to the final load. The final figure should represent the actual heat loss of the building as closely as it can be figured with the data available to the designer.

14.8 SUMMARY

Before a system can be designed to provide heating temperature control for any building, the heat loss of the building at some predetermined set of conditions must be figured. An accurate heat loss calculation enables a designer to determine the heat load for the building, size the heating equipment for the building, and determine how much air to circulate in the building. It allows the designer to size and locate the registers and size the air distribution and return systems. Without an accurately figured heat loss, these calculations are impossible.

To figure the loss, the designer must determine the outside temperature at which the building will be heated and the inside temperature to which it will be heated. The construction of the building must be considered so that proper factors can be selected to figure the amount of heat that will travel through the building components. All of these variables must be combined on a calculation sheet in such a way that the load can either be figured manually, or can be figured by using a computer system and computer heating/cooling load program.

The heat load calculation sheet should be one of two formats for manual load calculations, or a specialized form developed for use with a computer load program. In either case, the form should be one that is easy for a technician to use. One form used for manual load calculations assumes that the building is one large space or room. This is called a shell load. The other form used for manual load calculations assumes a room-by-room load is figured by determining individual room losses, and then adding all the room losses together for the total building load.

After the basic heat load is calculated, any additional heat losses or heat gains that might not have been included in the original load must be considered. The additional heat losses can include infiltration due to fireplaces, exhaust fans, duct work, or any other source of air leakage. The heat gains can include heat increases caused by the sun, by people in the building or by lighting in the building. The heat load should be adjusted accordingly.

The final heat loss figures should represent, as accurately as possible, the actual load of the building at the chosen design temperature. Accuracy in figuring the loads makes proper design of the complete system possible.

14.9 TERMS TO KNOW

heat load calculation forms
shell load
statistical data
perimeter distance
computer (cpu)
keyboard
disk drive
monitor
printer
solar gain

15.1 INTRODUCTION

After calculating the heat load, the heating technician must consider a number of other factors in order to select and size the heating equipment for a particular job. Proper consideration of each of these factors is necessary in order to choose the equipment best suited for a particular job.

The factors to consider in selecting and sizing the equipment are the heating capacity required, the physical arrangements possible, the fuel or energy available, the type of distribution system required, and the room available for location of the equipment in the building. Each of these factors is discussed in this chapter.

15.2 FACTORS TO CONSIDER

Several factors must be considered before the final selection of heating equipment for any building can be made. The major factors are

1. Heating capacity required
2. Type of system required
3. Fuel or energy available
4. Possible or potential equipment location
5. Local and national codes

The first factor to consider is the heating capacity required for heating a building. The heating unit selected must provide enough heat to match, or exceed, the calculated heat load. Enough heat must be available to maintain the desired temperature in a building at all times. The calculated heat load also determines the design of the distribution system. The system must be so designed that the registers, grilles, ductwork, and/or piping system will distribute the heat evenly throughout the building.

The two general types of systems considered here are the hydronic system and the warm air system. The size of the building, the location of spaces to be heated in relation to the equipment room, and the degree of temperature control desired are the general factors to be considered in deciding whether to use a hydronic system or a warm air system. A hydronic system usually is a better choice when the spaces to be heated are located at some distance from

CHAPTER

15

Selection and Sizing of Heating

the heating equipment. The warm air system usually makes more sense if the equipment room is located centrally and the heated spaces are more compact, If the building has shifting loads, such as groups of people that move around during the heating period or shifting solar loads, then the warm air system provides better control than the hydronic system.

One of the major decisions to be made in the selection process of a heating system is the fuel or energy to use. Cost and availability are the two main factors to consider. In some areas, certain fuels are more easily obtained than others; usually the cost for these fuels also is less. In many cases, the cost of transporting a fuel to a particular location is more than the difference in cost between two fuels. These differences have to be considered by the designer before the final selection of equipment.

In the design of a new building, space has to be provided for mechanical equipment, such as a furnace, blower, dampers, and ductwork or piping. The initial space arrangement is made by the architect designing the building, but the space left on the plans for the mechanical systems may not be adequate. The mechanical designer usually has to fit the mechanical equipment and system into whatever space has been provided. It is necessary, then, to select a system and equipment that will fit into the space available. Some types of equipment fit into tight spaces better than others. Water piping can be worked into floor spaces, walls, and ceiling spaces more easily than air ducts, because of the smaller sizes of pipes. But, water piping is very rigid and is not easily changed if flexibility in building space arrangement is important. A hydronic piping system cannot be adapted to changing conditions as easily as ductwork can. The entire building design has to be considered before the final selection of equipment is made.

Heating Capacity Required

It is necessary to select a heating unit with enough capacity to heat the building in which it is used. After the heat load has been calculated, two factors must be considered in determining whether a unit has the correct capacity to satisfy the load. First, the output capacity of the unit must be determined. Second, the need for extra capacity, or pickup, must be determined.

Output Capacity. Fossil fuel furnaces are listed in the manufacturers' catalogs with both input and output capacities. The **input capacity** is the amount of heat that will be given up by the fuel in the combustion process. The **output capacity** is the actual amount of heat available for heating purposes. Because of efficiency losses in the operation of a furnace, output in a conventional combustion type furnace is about 80 percent of the input rating. In a high-efficiency type furnace, with a secondary heat exchanger, the output may be as high as 93 percent to 97 percent. A designer must always use the output rating when sizing a furnace to match a heat load. The output of the unit selected should match or exceed the Btu/h required to heat a building at the design conditions.

Pickup. When a building is occupied continuously, 24 hours a day and every day of the week, the heating system will be in operation most of the time. The heating equipment has to be sized to offset the heat loss of the building on a continuous basis.

If building use is periodic (perhaps only one or two days a week), such as in a typical church or auditorium, the thermostat setting generally is set back to a lower temperature during unoccupied times. This thermostat setting is called **setback.** Traditional systems use a time clock and appropriate controls are used for achieving setback. A typical time clock is shown in Figure 15-1.

Most commercial buildings and many residential ones now use electronic control systems. Most of these systems include setback and recovery periods. Many of the commercial systems use electronic control systems that not only control the temperature but also ventilation, make up air, humidification, and control of other systems such as lighting.

Even in a setback system, heat loss during occupied periods of time is the same for the building as it is in an unoccupied time. The heating unit is sized to handle the total load, but it also is desirable to have extra heating capacity available to warm up the building more quickly when it comes off a setback period. This extra capacity is called **pickup.** If a building has been on setback for some time, everything in the building and the heating system cools off. Warming the building may take several hours after the thermostat is reset to the occupied temperature.

Input capacity*

Output capacity – The amount of heat available for heating a building after the input capacity is corrected for combustion and equipment efficiency.

Setback*

Pickup*

*Term defined in text.

FIGURE 15-1 _____

Time clock used for night setback on a heating system

Extra capacity in the heating unit will help warm the building faster in this circumstance.

So that the heating unit has this extra capacity, it should be sized from 15 percent to 50 percent larger than that required to offset the building heat loss. Building codes should be checked before deciding on how much extra capacity can be allowed for pickup. The actual amount of oversizing should be determined by how much time the temperature will be set back and how much difference there is between the on-time and off-time temperatures.

Types of Systems

The size of a building and the degree of temperature control required will determine whether a warm air system or a hydronic system should be used to heat the building. In a building in which the equipment is centrally located but heat is needed at widely separated parts of the building, a hydronic system may be indicated. If the building is more compact and ductwork can be run easily to each space to be heated, or if individual equipment spaces are available to each area, then a warm air system may be more practical.

Warm Air Systems.   Since the final desired result for any heating system is to heat the air in a building, a warm air heating system should be used whenever possible. This system is always better when the heating equipment can be centrally located and the ductwork does not have to be too long. Figure 15-2 shows a typical warm air furnace and ducts located in the basement of a home.

The use of a warm air system provides much more flexibility in the location of air outlets than does a hydronic system. Commercial buildings often require a system that can be changed if a tenant wants to rearrange spaces. In remodeling, it often is necessary to move the air outlets and return air registers. Because ductwork is much easier to relocate than rigid piping, the ultimate use of the building has to be considered by the designer in selecting the heating equipment.

Many commercial buildings are only one story high. Such buildings may be large in area and cover a lot of floor space, conditions not conducive to using a central warm air heating system. Instead, a **modular system** is practical. A modular system uses multiple heating units located at different strategic locations around the building, separate

Modular system*

FIGURE 15–2
Typical warm air furnace and ducts located in the basement of a house

from each other. The units may be located in separate equipment rooms or, in many cases, located on the roof. See Figure 15-3 for a typical heating and cooling unit as used on a roof.

In rooftop systems, the ductwork drops down through the roof through special openings. This type of system often is used for retail stores where large open spaces alternate with partitioned areas. Such modular systems can be either hydronic or warm air, but since the spaces can be kept relatively small, using warm air and running ducts to distribute the air makes more sense.

Hydronic Systems.   Whenever it is necessary to generate heat at one point in a building and distribute the heat long distances to the points of use, a hydronic system should be used. Water carries a much greater quantity of heat per volume of medium than does air. A hydronic system employs water pipes to carry the medium. The pipes also are smaller in size than are air ducts carrying a similar

FIGURE 15–3
Rooftop heating and air-conditioning unit showing ductwork that goes through the roof

amount of heat. In a large building or a multistoried one, too much space may be required for ductwork, or in some cases, the construction of the building will not allow space for the ductwork.

High-rise buildings and campus-style buildings scattered around a central heating plant are good examples of the type of buildings in which a hydronic system can be used to good advantage.

In a hydronic system, a boiler in a central equipment room generates hot water. Pumps circulate that water to its point of use as shown in the diagram in Figure 15-4.

The designer must be sure that the building has appropriate spaces for the equipment and that the construction of the building will allow the distribution piping to be run.

In the spaces to be heated, terminal devices are employed to extract the heat from the water. Heated air is then used to heat the spaces. The designer should consider what type of terminal devices to use so the total system design will accommodate them. One type of terminal unit is shown in Figure 15-5.

Fuel Types

A major decision to be made in the selection of heating equipment concerns fuel choice. In this discussion, electric-

FIGURE 15-4

Schematic drawing of a hydronic heating system

FIGURE 15–5

Convector unit used as a terminal unit in a hydronic system

ity will be considered a fuel. Units that use any of the fossil fuels or electricity are available in either the warm air system or the hydronic system. The choice of which fuel to use normally depends upon the availability and cost of fuels in the area where the particular building is located. In the past, when such factors were reasonably stable, only the long-term effects of changes had to be considered. But now, both cost and availability change so rapidly that it is necessary to select the fuel with great care, and also to design a system so that the fuel can be changed in the future if availability or cost factors change.

Fossil Fuels.   In most populated areas, natural gas pipelines are common. If natural gas is available, it seems to be the most stable of all the fuels used in the United States at the present time in respect to price and availability.

Oil is an excellent heating fuel, and many of the heating plants in the United States are set up to use it. Use of oil for a heating fuel in the United States far exceeds the domestic supply, however. The cost per Btu/h of heating with oil now exceeds that for either gas or electricity in most places in the United States. The future use of oil as a heating fuel probably will be limited.

Many large industrial plants are purposely located on or near sources of coal. For such buildings, the choice of which fuel to use is simple. Coal is a good source of heat. When it can be used close to its source in order to minimize the cost of transportation, and when the contamination from the products of combustion can be controlled, so they are not detrimental to the environment, coal is an excellent choice for heating.

Electric Energy.   Since electricity is available in virtually all buildings, this form of energy is usually a good choice as a source for heating. In some areas, the cost of electricity may be greater per Btu/h than gas, oil, or coal, but since it is nearly always available at a building site it can be used for heating when other fuels may not be available. In some cases, electric utility companies limit the use of electric power because the demand exceeds the supply. The designer should always check with the local supplier of any fuel or energy before deciding on any particular one to use for heating a building.

Equipment Location

Gas, oil, and coal need combustion for conversion to heat. This requirement makes it necessary to provide a way to get fuel to the equipment, to provide combustion air, and to provide a chimney, flue, or vent to get rid of the products of combustion. Building design has to include space for fuel lines or bunkers (in the case of coal), a chimney, and combustion air ducts to and from the outside of the building. Space has to be provided for the distribution ducts or piping. The air distribution system also requires fresh air, so ducts must be provided to bring this air into the area where the heating equipment will be located.

Fuel Availability.   Gas or oil can be piped to the equipment room from outside meters or tanks. Because only a small amount of space is required for running the lines, the lines normally do not affect the location of the equipment room. A tank and piping for an oil-fired furnace are shown in Figure 15-6.

The electric wiring circuits to an electric heating system usually can be run wherever necessary in a building, and control panels can be located where convenient. The

Vent pipe

Filler pipe with cover

Supply pipe to furnace

Basement wall

Supply pipe to bottom of tank

Underground tank

FIGURE 15-6 _____
Buried tank and the piping installation for an oil-fired heating system

wiring circuit chases are small, so the equipment room location is not significantly affected by them.

If coal is to be used as the fuel, the equipment room should be located adjacent to the coal storage area so stoking the furnace is simplified. The heating designer and the building architect need to coordinate their work to ensure proper location of these areas.

Venting. When combustion equipment is being used, the products of combustion must be vented to the outside of the building. A vent or chimney has to be provided from inside the equipment room to the outside of the building. The vent, for conventional furnaces, should run vertically up through the building, and will need to be sized properly for the equipment used. If high efficiency furnaces are used with power burners or draft-inducing fans, the vents are usually much smaller, and may be run horizontally or vertically out of the building. The length of the vent run becomes critical if it is run horizontally.

The design of the building has to be such that the chimney can be run up from the equipment room through the rooms above to the outside of the building as shown in Figure 15-7. Since rooms and partitions on the second floor do not necessarily line up with those on the first floor, this positioning is not always easy to do. The chimney or vent normally is enclosed in an opening in the partitions or at least in a closet. If the partitions do not line up, special arrangements may have to be made to accommodate the chimney on the second, or any other, floors. It usually is necessary to locate the equipment where it is convenient to the chimney. The designer should keep this problem in mind when selecting the equipment for any job.

Ductwork or Piping. In warm air systems, supply ducts run from the heating equipment to the different rooms in the building. Return ducts bring the air back to the equipment. Space has to be available for these ducts. It is most desirable to have the ducts concealed in the construction of the building. Normally they are installed in ceiling spaces, furred-down sections in halls, crawlspaces, or in special spaces called **chases** that are provided in the building frame. The designer should make sure that such spaces are available before selecting the heating equipment. Figure 15-8 shows the ductwork for one type of system, a perime-

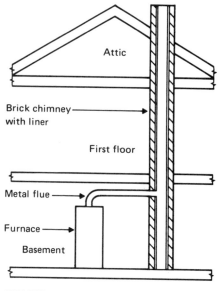

FIGURE 15–7
Chimney in a multistory building

Chase*

FIGURE 15–8 _____
Diagram of ductwork for perimeter loop system used with a warm air system

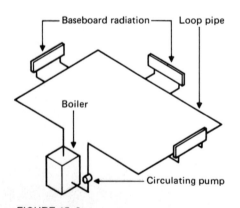

FIGURE 15–9 _____
Diagram of piping for a simple loop system used with a hydronic system

ter loop system, which normally would be located under the floor. Other types of duct systems are shown in Chapter 16.

A hydronic system utilizes supply and return piping to transport heat. Water pipe is much smaller than air duct for a given amount of heat, so it is easier to find room for water pipes than it is for air ducts. A piping system, however, is much more rigid and inflexible in design than are air ducts. Changing a piping system is more difficult than changing a duct system. Figure 15-9 shows the piping for a simple loop system.

Access.   Space should be allowed around mechanical equipment to provide for maintenance and service. Manufacturers' installation manuals show the proper amount of clearance to allow on each side of the equipment. The designer must always provide for these allowances when laying out the equipment room.

Proper access should also be planned to the equipment room. If the room is located above or below the main traffic area in a building, stairs or access ladders should be provided. Space should also be allowed for eventual replacement of parts or even of the entire unit. Doors, stairways, and lifting facilities should be large enough to accommodate these parts. The designer should ensure that such arrangements can be made for the equipment chosen for the job.

Combustion Air.   A designer must provide for a constant supply of combustion air to the equipment room whenever equipment is used that employs combustion for heating. Combustion air should be brought in from the outside of the building. See Figure 15-10. If the equipment room is located on an outside wall, part of the wall may be **louvered** for combustion air and fresh air. The type of fuel used will determine the size of the opening required, but the openings and ducts should always be large enough to provide at least 50 percent more air than that required for complete combustion. As a general rule, openings with at least 1 sq in. per 1000 Btu/h of furnace input should be provided at the floor line and ceiling line of the furnace room.

If the equipment room is located in the central part of a building, ducts usually have to be used to bring the air into the space. If such is the case, the designer should make sure there is room for the ducts to run through the building framing or that duct chases are provided.

When electric heat is used, combustion air is not required since no combustion takes place.

Ventilation Air.   To keep the atmosphere in the building healthy and comfortable, ventilation air must be introduced into the occupied spaces. The ventilation air is introduced into the return **plenum** of the heating system. The air is thus processed through the filters and the furnace and is introduced into the spaces with the supply air. The same amount of return air has to be exhausted or vented so that the pressure in the building remains constant.

The amount of air introduced as ventilation air is determined by the number of people who will be in the building, by the amount of contamination of the air in the building, and by the cubical content of the building. This topic is covered later in the chapter.

Ventilation air, like combustion air, is brought into the equipment room through ducting or louvers. Unlike combustion air, however, it is ducted directly to the return air side of the air mover. Therefore, the designer has to make sure there is room for the ducts in the building design.

FIGURE 15-10
Providing ventilation and combustion air for equipment room

15.3 EQUIPMENT SIZING

After a heating unit has been selected to match the heat load, and to fit into the building design, the amount of air or water to circulate in the system should be determined. To do this calculation, the Btu/h output of the heating unit is converted to cfm of air for a circulated air system, or to gpm of water for a hydronic system. The heat itself is carried by the air or water; hence the need for the conversion.

Air Circulation.   For a warm air system, the cfm of air circulated generally is based on the temperature rise through the furnace used, along with the air changes in the building space. Temperature rise through the furnace is a variable selected at random within certain limits. The rise is normally figured at 80°F to 100°F for fossil fuel units, 50°F for electric heat units, and approximately 20°F for a heat pump without auxiliary heat.

Louver – An opening that allows air to enter but that has overlapping outward-sloping vanes that prevent rain from entering.

Plenum – The box on the inlet or outlet of the furnace that connects the ductwork to the furnace.

Btu/h to cfm

To find the cfm of air to circulate for the Btu/h output of a funace, the formula for sensible heat can be used. Although there always is some moisture in the air being heated the effect of that moisture is not noticeable on a heating application. The sensible heat formula is

$$cfm = \frac{Btu/h}{TD \ 1.08}$$

where

cfm	= amount of air to circulate
Btu/h	= amount of heat involved
TD	= temperature difference (rise or drop)
1.08	= specific heat of standard air × the specific weight × 60 min in 1 hr

Note: The figure 1.08 is mathematically correct. It is derived by using the values for standard air. In present-day practice the figure 1.1 is often used. This is more common in air conditioning applications then in heating applications. Either figure will give results close enough for actual practice.

EXAMPLE 15-1

Find the cfm for a gas or oil-fired furnace of 112,000 Btu/h output (the output for a 140,000 Btu/h input furnace) and an 80°F temperature rise.

Solution

$$cfm = \frac{Btu/h}{TD \times 1.08}$$

$$= \frac{112,000}{80 \times 1.08}$$

$$= 1297.30$$

The amount of air that the system should be designed for is 1297.30 cfm.

Air Change. A second way to determine the total amount of air that should be circulated in a building is to plan on changing the air in the building a given number of times per hour. When the air change per hour basis is used to evaluate the amount of air needed, the following formula can be used:

$$\text{cfm} = \frac{V \times AC}{60}$$

where:

cfm = amount of air to circulate
V   = volume of the building
AC = number of air changes per hour
60   = min in an hr

The building volume, in cu ft, is multiplied by the number of air changes desired per hour. The result is divided by 60 min. For the total air to be circulated, anywhere from 4 to 8 air changes per hour may be used. If the formula is used to determine ventilation air only, then 1/2 to 2 air changes per hour should be used, the variation depending upon how tightly the building is constructed.

EXAMPLE 15-2

The floor plan in Figure 14-5 in Chapter 14 shows a residence 48 ft × 30 ft, with an 8 ft ceiling height. Find the amount of air to circulate for 6 air changes per hour in this building.

Solution

First, find the volume of the building:

$$\begin{aligned} V &= l \times w \times h \\ &= 48 \times 30 \times 8 \\ &= 11{,}520 \end{aligned}$$

The volume of the building is 11,520 cu ft. Now, use the air change formula:

$$\begin{aligned} \text{cfm} &= \frac{V \times AC}{60} \\ &= \frac{11{,}520 \times 6}{60} \\ &= 1152 \end{aligned}$$

TABLE 15–1 ———————————————
Ventilation Rate Table

Smoking	cfm per person
Light	5–10
Medium	10–25
Heavy	25–40

Note: Figures are cfm of air per person to be brought into the return air system from outside the building.

The amount of air to circulate through this building for 6 air changes per hour is 1152 cfm.

After the amount of air to circulate has been determined by either of the above methods, the temperature rise of the air through the heating unit with that amount of air should be checked. In some cases, the amount of air will need to be adjusted to provide a proper temperature rise through the furnace.

People Load. If the people load is high in a building, ventilation air should be introduced into the circulated air to keep the air in the building from becoming stale. The amount of ventilation air that should be circulated can be calculated on the basis of the number of people in the building. A general rule of thumb is that the more smoking in the building, the more ventilation air will be required to keep the air fresh. If the smoking is light, then approximately 5 to 10 cfm per person of ventilation air should be introduced to the system. If the amount of smoking is medium, from 10 to 25 cfm should be brought in. If the smoking is heavy, from 25 to 40 cfm per person will be required. Table 15-1 is a ventilation table giving guidelines to use for ventilation rates.

There are some other obvious situations in which ventilation air is required. Among these are buildings or building spaces that have high concentrations of chemicals that may be either objectionable or dangerous due to their nature. In these applications the ventilation air usually replaces exhausted air. Proper sizing of the ventilation and exhaust systems in these applications depends upon the size of the spaces involved, and the amount of contamination.

Remember: ventilation air is to be brought in from outside and circulated with the return air. Ventilation air cfm is not the same as the total air to be circulated. A general rule is that the ventilation air will be from one-tenth to one-third of the total air circulated.

Btu/h to gpm

When a hydronic system is used, Btu/h must be converted to gpm. This conversion is done by dividing the Btu/h by the temperature difference multiplied by a factor of 500. The temperature difference in this case is for the temperature drop of the water as it goes through the terminal units in the system. The formula to find gpm is as follows:

People load – The actual number of people in a building.

$$gpm = \frac{Btu/h}{TD \times 500}$$

where:

gpm	= flow rate of the water
Btu/h	= heat loss of the building
TD	= temperature difference (rise or drop)
500	= approximate factor derived from the weight of water per gal × the specific heat of water × 60 min per hr

EXAMPLE 15-3

If a hydronic system is used to heat the building in Example 15-1, with a heat loss of 112,000 Btu/h and a TD of 20°F (about right for a small hydronic system), find the gpm.

Solution

$$gpm = \frac{Btu/h}{TD\ 500}$$

$$= \frac{112,000}{20°F\ 500}$$

$$= 11.2$$

To carry the 112,000 Btu/h of heat, with a TD of 20°F, 11.2 gpm of water must be circulated in the system.

gpm and Terminal Units.   In a hydronic system, the heat must be extracted from the water at the point of use by some device. This device is called the terminal unit of the system.

The amount of heat extracted by a terminal unit is found by using another form of the formula shown in Example 15-3. This new formula is:

$$Btu/h = gpm \times TD \times 500$$

where:

Btu/h	= heat extracted
gpm	= flow rate of water
TD	= temperature difference
500	= approximate factor derived from the weight of the water per gal × the specific heat of water × 60 min per hr

The heat extracted from the water by the terminal units is distributed in the building by different methods, depending upon the type of terminal units used. The different types of terminal units and methods of distribution are covered in Chapters 19 and 20.

Local and national building codes must be considered in all applications of heating systems. Safety requirements, and consideration of energy use have made many code requirements mandatory. New and amended code requirements should be considered on every installation.

15.4 SUMMARY

After an accurate heat load of a building has been calculated, the next step in designing a system is to select the heating equipment. Selecting the heating equipment that is best for a building is based on whether a warm air system or a hydronic system should be used, what fuel or energy is available, and where the equipment can be located in the building.

Since the ultimate purpose of most heating systems is to heat the air in a space, the warm air system is most commonly used. If the heating equipment has to be located some distance from the point of use, however, a hydronic system may be more practical.

Heating units that use any of the fossil fuels or electricity are available in either warm air systems or hydronic systems. The choice of fuel or electricity is a matter of availability, price, and convenience of use in the particular building to be heated.

Combustion equipment requires combustion air, venting of the fuel gases, and piping and/or storage of the fuel. Electric units do not need to be vented, but they require wiring and controls for the electric power. All factors have to be considered when selecting a system or equipment.

The quantity of air or water to be circulated has to be considered when selecting heating equipment. The cfm of air is figured on the basis of the amount of heat to be distributed and the temperature rise of that air through the furnace. The gpm of water in a hydronic system is based on the amount of heat to be distributed and the temperature drop of that water through the system. The cfm or gpm are calculated by use of standard formulas.

When a hydronic system is used, the heat has to be extracted from the water at the point of use and transferred to the air for the actual heating of the space. The final act in designing a hydronic system is to figure the cfm of air required at this point of exchange.

15.5 TERMS TO KNOW

input capacity
output capacity
setback
pickup
modular system
chase
louver
plenum

CHAPTER
16
Warm Air Distribution and Duct Sizing

16.1 INTRODUCTION

In a warm air heating system, the heat is generated by a furnace at one point in the building and distributed to the various parts of the building by air. A duct system carries the air where heat is wanted. The ductwork is made up of metal or fiberglass ducts laid out in a pattern that best distributes the air.

The general design of the duct system is determined by the construction used for a particular building. If the building is constructed on a concrete slab on ground, one type of system is used. If the building has a crawl space or a basement, another type of system is required. A designer must decide which system is best for each job.

After the type of system is decided upon, the final step is to size the ductwork so the proper quantity of air is delivered to each part of the building.

16.2 TYPES OF DUCT SYSTEMS

A warm air heating distribution system includes the supply and return duct systems. The supply duct system usually is the more complex of the two. Three general types of distribution systems are used. They are

1. Perimeter loop system
2. Radial system
3. Trunk and branch system

These designations refer to the system's **configuration** and are normally the same whether they are installed under the floor or overhead. There are many variations of these basic configurations. However, even those configurations used in large buildings are combinations or variations of the basic systems.

Perimeter Loop System

A **perimeter loop system** is one in which the ductwork runs around the perimeter of a building in a complete loop. The supply registers are located on the loop, and **branch ducts** carry air to the loop from the supply air plenum.

Configuration – Shape or pattern of a design.

Perimeter loop system*

Branch ducts – Ducts used to feed air into a loop or take air out of a trunk for distribution to registers.
Loop – The duct that goes around the perimeter of a building, is fed by branches, and has registers located on it.

Radial system – A duct system that has individual branches radiating out from the supply plenum, each supplying air to a register.

*Term defined in text.

This system is most commonly used when the building has a concrete floor poured directly on the ground. The outside loop of the duct runs around the perimeter of the building under the edge of the concrete floor slab. Figure 16-1 illustrates a perimeter loop system.

When the ducts for a perimeter loop system are sized, the loop pipe is divided into sections with registers in each section having approximately the same cfm requirements. The loop is then sized to provide the cfm required for the section needing the greatest cfm. The loop pipe is the same size all around. The feeder pipes are sized to provide the cfm required for the sections in the area covered by the feeder.

Radial System

A **radial duct system** is one in which branches run out radially from the warm air plenum to each register, as shown in Figure 16-2.

A radial system normally is used when ductwork is run in an attic or a crawl space of a building. This system should be used only where head room or walkways will

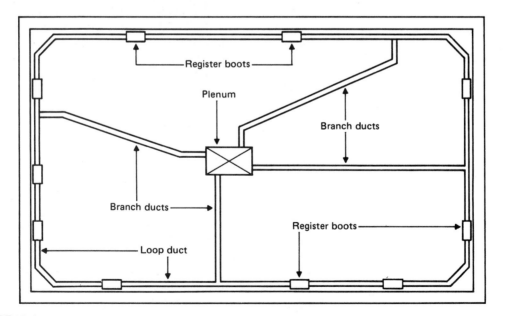

FIGURE 16–1
Perimeter loop duct system

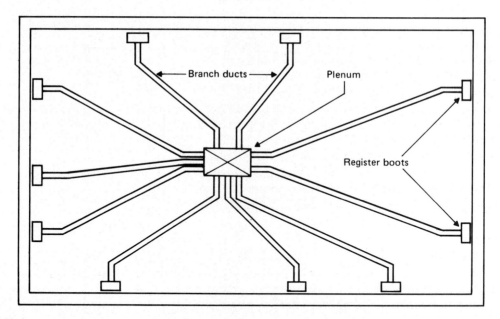

FIGURE 16–2
Radial duct system

not be affected by the ducts. It is especially adapted to buildings where the heating unit is located approximately in the center of the building, and the ductwork can run directly from the supply plenum at the furnace to each supply boot. Each branch duct in a radial system is sized for the amount of air required by the register located on that branch.

Trunk and Branch System

A **trunk and branch system** is one in which a main trunk, or trunks, extend out from the plenum, and branches run from the trunk to each register. Figure 16-3 is a diagram of a typical trunk and branch system.

This system often is used on systems where the branches can be run either up in the spaces between floor joists, or in the spaces between the ceiling joists. The trunk that feeds the branches is run along a beam or supporting partition, where it can be enclosed as the building is finished. A good example of this type of system is a duct system installed in a finished basement. The ductwork is installed in the ceiling and then covered up by the ceiling

Trunk and branch system – A duct system with a main trunk that supplies air to separate branches, which in turn supply individual registers.

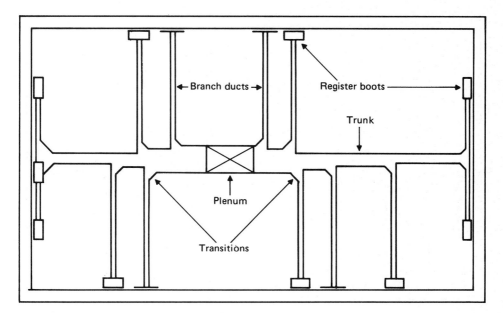

FIGURE 16–3
Trunk and branch duct system

finish material when the basement is finished.

The main components of a trunk and branch system are the main trunk and the branches that lead off it. The trunk is sized to carry the air needed for the part of the building that it serves, and each branch is sized to carry the air required by the registers they serve.

There are two types of trunk and branch systems: (1) an extended plenum system and (2) a graduated trunk system.

Extended Plenum Trunk and Branch System. In an extended plenum trunk and branch system, the main trunk is kept the same size for its entire length, and each branch is a uniform size for its entire length. This system is illustrated in Figure 16-4. An extended plenum system should be used only on small jobs, where the length of the trunk is minimal. This system is difficult to **balance.** If the main trunk is not short, some branches will get more air than others.

Graduated Trunk System. In the **graduated trunk system,** the size of the main duct is reduced as the quantity of air is reduced at each **branch take-off.** A branch take-off is the fit-

Extended plenum trunk and branch system*

Balance – Adjusting an air supply system so the proper amount of air will be delivered by each register in the system.

Graduated trunk system*

Branch take-off – The fitting between the main duct and a branch duct.

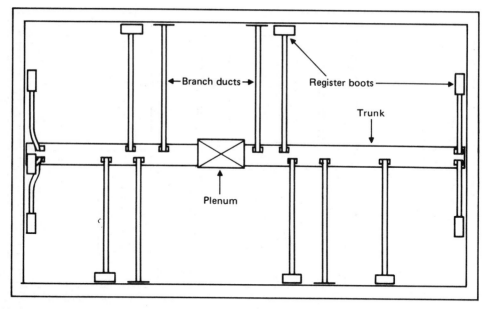

FIGURE 16-4

Extended plenum, trunk, and branch duct system

Static pressure – The pressure of a moving stream of air, measured at right angles to the direction of airflow.

ting where the branch fits onto the trunk. This system is designed so air for the branch flows smoothly from the trunk into the branch. This system is illustrated in Figure 16-3. Because of the duct reductions, the velocity of air remains the same throughout the trunk, and consequently so does the **static pressure**. The system is shelf-balancing. The proper amount of air will flow out of each branch take-off, regardless of how far the take-off is from the supply plenum.

16.3 LAYING OUT THE SYSTEM

After the type of air distribution system is decided upon, the next step in designing is to transfer all of the data concerning the job onto a plan of the building. The designer draws, or obtains an outline plan of the foundation or basement and a plan for each floor of the building from the building designer or architect. These plans should show the walls, partitions, windows, doors, cabinets, and any other details that will influence the location or size or registers, ducts, or other parts of the heating system.

FIGURE 16-5
Floor plan of a typical residence used to prepare a duct system layout

Special note should be taken of construction details in the vicinity of the equipment room. Supporting columns or beams that run where the designer needs to place ductwork should be especially noted. Such major parts of the construction cannot be moved, and the designer should avoid them in showing the location for the ducts. Piping from the kitchen or bathrooms is another example of construction details that are necessary to the building. The designer should avoid the areas containing piping when designing ductwork. If the plans are on tracing paper, details can be transferred from one plan to another easily.

The plans should be drawn to as large a scale as is practical, depending upon the size of the building, but no smaller than 1/8 in. to a foot for larger buildings and 1/4 in. to the foot for smaller buildings. Figure 16-5 is an example of such a floor plan for a typical residence.

Design Data

Using the floor plans, the designer should make a note in each room of the Btu/h and cfm required in the room. The Btu/h and cfm are taken from the calculated heat load for the building. This information has been added to the plan in Figure 16-5.

After the Btu/h and cfm have been recorded on the plan, a tentative selection of supply registers is made. The registers are selected to provide the required amount of air to each room. This selection should be made so the registers will be compatible with the type of system decided upon.

Register and Grille Location

Using the floor plan, the designer marks the location of the registers. The plan in Figure 16-5 has the registers indicated with the proper symbols.

The designer uses the same procedure to select and show the location of the return air grilles on the plan. See Figure 16-5. Finally, the designer determines the correct sizes for the registers and grilles and applies this information to the plan. See Figure 16-5.

After the supply registers and return air grilles are drawn on the floor plan, a tracing of the foundation plan, or whatever floor the ductwork is going to be located on, is placed over the floor plan. The location of the registers and grilles is shown on this overlay. Figure 16-6 is an example of this drawing.

FIGURE 16–6

Overlay of the foundation or basement plan with the register and grille locations applied

Ductwork Layout

With the register locations marked on the plans, the duct-work can be tentatively drawn in. It is now possible to visualize the layout of the system, to see if there is any conflict with other components of the building construction, such as posts, beams, plumbing, or electrical details. Figure 16-7 shows the layout of the supply and return ducts drawn on the foundation overlay.

If there is no apparent conflict, the ductwork can be drawn in, either in single line or full scale. A single line drawing is one that shows the walls, partitions, and other major parts of the building by the use of a single line for each. Thickness of materials is not indicated. The plan may be drawn approximately to scale, but individual parts are not exactly to scale. A full-scale drawing shows all parts in double lines so the thickness of each part is indicated on the plan, as well as the location in relation to other parts. The ductwork should be drawn on the plan as it will actually be placed in construction. The location of the heating unit or equipment room must also be taken into consideration when locating the duct system. After the main trunks are located on the plan for both the supply and return air systems, the branch ducts from the trunks to the register boots are drawn in, taking the same precautions pertaining

FIGURE 16-7
Tentative layout of the supply and return ducts made on the foundation overlay

Performance standards – A written description of the expected results of the operation of a system, usually giving temperature, quality, velocity of air, and any other criteria necessary to describe desired performance.

to building components as used for the main trunks. Ductwork for make-up air and combustion air, if required, should also be drawn in at this time.

If the ductwork take-offs and returns to the equipment are complex, it will be difficult to show them on the floor plan. Larger-scale drawings may be needed for the equipment room. The drawings should show the supply and return plenums and the connections of the ductwork. Plan views, sections, and elevations can be used to show the details necessary.

Equipment List and Specifications

The equipment should be listed on the plans so that estimators and installers know exactly what equipment is required for the job. The equipment can be shown on a schedule on the heating plan, as well as in the specifications for the job. Table 16-1 shows an equipment list for a typical job. The equipment list should include registers, grilles, dampers, controls, and any other equipment or accessories required for the heating system. Notes of any special requirements should be made on the plans to call attention to areas of special concern. Location of the thermostat should be shown on the plan.

A description of the equipment and accessories should also be made in the specifications for the job. The specifications should include the **performance standards** for the equipment and all of the components of the system, as well as physical or mechanical standards for the installation. The specifications and plans become part of the contract documents for the job and are extremely important in ensuring that the job is installed as the designer intends.

16.4 SIZING THE DUCTS

After the duct system is drawn in on the plan, the size of the ducts at various locations throughout the system has to be determined and the sizes marked on the plan. All transitions should be noted where the duct changes from one size to another. The dimensions of the ducts, with all drops, transitions, elbows, and other components of the system, are marked on the plan. It is important that all

TABLE 16–1 ───────────────────────────────────
Typical Equipment List

Equipment	Specifications
Furnace	Shall be an upflow, gas-fired unit with 80,000 Btu/h output, and shall be capable of heating the home to 75°F when the outside temperature is 15°F. The blower shall be capable of delivering 1000 cfm of air at a total static pressure of .50 in. wg. The furnace shall be equipped with electric ignition and have a ceramic-coated heat exchanger.
Ductwork	Shall be of galvanized iron, installed in a proper manner and according to latest standards. Air turns shall be used in all elbows, and flexible connections shall be used at the connection to the plenums. Balancing dampers shall be installed in all branches and in the fresh air intake duct.
Registers and grilles	Shall be of steel construction as noted in the specifications, shall be painted with a rust-preventing undercoat, and shall match the room surfaces.
Air cleaner system	Shall include an electronic air cleaner on the inlet side of the furnace, which will be rated to handle the total cfm of the system with a cleaning efficiency of 80% by standard dustspot test at that cfm.
Humidifier	Shall be a spray-type humidifier installed in the supply duct to provide enough moisture to maintain a 50% Rh in the building at all conditions, and shall be controlled by a humidistat placed in the return duct.

dimensions are given so the estimator can make an accurate take-off for the material and estimate the labor that will be required to install the job. A **take-off** is an accurate list of materials needed for the construction of a building or part of a building. In the case of the ductwork system, the take-off should show all equipment and materials necessary for the installation of the ductwork system. The installer will also use the plans for making up the ductwork. Details of the return air system should be shown on the plan.

The ductwork must be carefully sized to ensure that the correct amount of air gets to each room or space to provide proper heating. Any adjustment in the calculations of the cfm to be delivered to the rooms should be made before the duct is sized.

Two methods of duct sizing will be described in this text: (1) the equal velocity method and (2) the equal friction method. These methods are the basic ways to size ductwork. All other methods are an extension of them.

Equal Velocity Method

Air noise in a duct is directly related to the construction of the duct and the velocity of the air. The equal velocity method of duct sizing is used when a designer wants to maintain a given velocity of air in each section of the system to control the sound level and still use as small a duct as possible for economic reasons. The velocity of the air in the ducts is a function of the size of the duct and the quantity of air flowing in it. This relationship is best shown by the following formula:

$$V = \frac{Q}{A}$$

where

   V = velocity of the air in ft per min (fpm)
   Q = quantity of air in cfm
   A = area of the duct in sq ft

EXAMPLE 16-1

If a section of ductwork is 12 in. wide and 9 in. high, and 600 cfm of air to pass through it, what will be the velocity of that air?

Solution

To start with, the dimensions of the duct are given in inches (this is the correct way to size duct). The area of the duct in square inches is:

$$A = w \times l$$

where:

A = area of the duct, in the same terms as w and l

w = width of the duct

l = length, or in this case height

So, for this example:

A = 12 × 9 = 108

The area 108 sq in has to be changed to sq ft to use in the velocity formula. Since there are 144 sq in., in a sq ft, then 108 sq in. divided by 144 sq in. gives the area in sq ft.

$$A = \frac{108}{144} = 0.75$$

The area of the 9 × 12 duct in sq ft is .75. The area can be used in the velocity formula, as follows:

$$V = \frac{Q}{A} = \frac{600}{0.75} = 800$$

The velocity of the air in the duct is 800 fpm.

When the equal velocity method of duct sizing is used, a velocity table is used to find acceptable velocities for each application. Table 16-2 is an example of a velocity table. This table gives the maximum allowable velocity for the air in the ducts, for various types of installations. Reference to the table will show that the velocity of the air in the main trunks can be higher than that in the branches, and that the acceptable sound level in various types of buildings is also considered.

To use the table, find the type of building use and select the acceptable, or maximum, velocity in the formula. The quantity of air required through each part of the duct in the system is taken from the layout and the data sheets already prepared. The allowable velocity is taken from the table. The duct size is figured by use of the following formula:

$$A = \frac{Q}{V}$$

TABLE 16–2
Recommended Duct Velocities for Various Applications

Applications	Residences	Schools, Theaters, Public Buildings	Industrial Buildings
	Velocities for Total Face Area		
Outdoor air intakes	500 (800)	500 (900)	500 (1200)
Filters	250 (300)	300 (350)	350 (350)
Heating coils	450 (500)	500 (600)	600 (700)
	Velocities for Net Free Area		
Air washers	500 (500)	500 (500)	500 (500)
Suction connections	700 (900)	800 (1000)	1000 (1400)
Fan outlets	1000–1600 (1700)	1300–2000 (1500–2200)	1600–2400 (1700–2800)
Main ducts	700–900 (800–1200)	1000–1300 (1100–1600)	1200–1800 (1300–2200)
Branch ducts	600 (700–1000)	600–900 (800–1300)	800–1000 (1000–1800)
Branch risers	500 (650–800)	600–700 (800–1200)	800 (1000–1600)

Note: Maximum duct velocities appear in parentheses.

when:

A = area of the duct in sq ft
Q = cfm of air
V = velocity of air in fpm

EXAMPLE 16-2

If a branch duct is required to carry 250 cfm of air, and the velocity chart shows the allowable velocity for that part of the system to be 600 fpm, find the area of the duct.

Solution

$$A = \frac{Q}{V}$$

$$A = \frac{250}{600}$$

$$= 0.416 \times 144 = 60$$

The area of the duct is 0.416 sq ft. Multiplying that by 144 sq in., in a sq ft, gives 60 sq in.

The actual dimensions for square, rectangular, or round duct can be found from the area mathematically.

When rectangular duct is used, the best practice is to make the duct height not less than 8 in. and to limit the width to not more than three times the height. When a 24 in. × 8 in. duct is not large enough to accommodate the cfm, then the depth of the duct should be increased to 10 in., and so on, keeping the width at no more than three times the height. When sizing ducts, the width or horizontal dimension is always given first, the height or vertical dimension is given second.

FIGURE 16–8 _____
Duct system layout showing the duct sizing and location of the transition

Equal Friction Method

The equal friction method of sizing ducts is based on the fact that as air moves through a duct, it causes a certain amount of resistance to the flow. This resistance is a function of the amount of surface exposed and the quantity of air flowing.

For air to move through a duct, there must be a pressure difference between the inlet and outlet of the duct. The blower, or air mover, exerts a higher pressure on the air at the inlet end of the duct than exists at the outlet. This pressure difference makes the air move. The amount of difference indicates the amount of resistance in the duct. The pressure is measured with a device called a manometer. Figure 16-8 shows a manometer of the type used for measuring pressure in ducts.

The pressure, called the static pressure, is read by taking a manometer reading through a hole in the wall of the duct at a 90° angle to the airstream. The measurement is in inches of water. A more complete explanation of static pressure and the use of a manometer is given in the next chapter. Friction in the duct causes a reduction in the static pressure as the air moves through the duct. The static pressure difference between the inlet and the outlet of the duct can be used to indicate the amount of friction in the duct. Since friction, and consequently **static pressure drop,** are functions of cfm and duct size, it is possible to use static pressure drop for a given length of ducts and cfm to

Static pressure drop – The difference between the static pressure readings at two points in a duct system, which indicates the effect of resistance to airflow.

determine duct size. Charts, called equal friction charts, have been developed that show the relationship of static pressure drop, cfm, and duct size. Such a chart is found in Table A-8 in the Appendix. To use the equal friction chart to find the size of duct used to deliver a given cfm of air, find the static pressure (SP) drop per 100 ft of duct on a vertical line on the chart and the cfm on a horizontal line. Where the two intersect, the duct size is shown on a diagonal line coming down from the top right of the chart. The duct size is given for round duct.

To size the duct by the equal friction method, select the amount of friction desired per 100 ft of straight duct and use that amount as the base for the calculations. For residential and smaller commercial or industrial systems, a static pressure drop of 0.10 in. wg is about right for each 100 ft of duct. The designer, using the cfm required for a given section of duct and the static pressure drop per 100 ft of duct, uses the equal friction chart to find the duct size required.

EXAMPLE 16-3

In a given application if 250 cfm of air is required and a static pressure drop of 0.10 in. per 100 ft of duct is to be used, find the size duct required.

Solution

The 250 cfm line is found on the horizontal scale of the equal friction chart. The static pressure drop of 0.10 in. is found on the vertical scales. Where the two cross, the duct size is found on the diagonal line coming down from the top right of the chart. In this example, the size is found to be just slightly larger than 8 in. in diameter. An 8 in. duct would be used.

If a rectangular duct is required, rather than a round one, then a round-to-rectangular conversion chart should be used to find the equivalent size of rectangular duct. A rectangular duct equivalent chart is found in Table A-9 in the Appendix. When the equivalent chart is used in Example 16-3, a 7 in. × 8 in. duct would be used in place of the 8 in. round.

Showing the Duct Sizing on the Plan

After the duct sizes have been determined, the next step is

FIGURE 16–9
Duct system layout showing the duct sizing and location of the transition

to record them on the drawing already made of the system. The best way is to start sizing the duct from the extreme outer end of the branches or trunks. Work back toward the air mover, sizing each duct in turn. Increase the size where each branch takes off, and letter in the size on the layout. All transitions, or changes in duct sizing, should be indicated on the plan. Figure 16-9 shows the duct sizing noted on the plan.

This procedure should be used for each branch of trunk that leads off from the plenum at the unit. After each branch is sized, the quantity of air that the ducts have been sized to carry should be rechecked to be sure that the figure equals the output of the blower.

After the supply air system has been sized, the same procedure can be used to size the return air system for the job. The design of the return air system is just as important as the supply side of the system.

16.5 TOTAL SYSTEM FRICTION

After a duct system is sized, a designer must make sure that the heating unit used has a blower capable of delivering the cfm required against the friction in the system. This requires calculating the total friction in the duct system. When the equal friction method of duct sizing is used, this operation is easy. Since the duct sizing is based on a given amount of friction per 100 ft of duct, it is necessary only to calculate the equivalent length of duct in the longest run for the supply side of the system, and to make the same calculation for the return side. The total equivalent lengths include the actual lengths of the straight sections of duct plus equivalent lengths for any fittings in the sections. Equivalent lengths for fittings are found in appropriate tables. The equivalent lengths for supply side and the return side of the system are added together to get the total equivalent length for the system. The sum is divided by 100 (since the original figure was for friction per 100 ft of duct). This result is multiplied times the static pressure drop per 100 ft of duct used. The resulting figure is the total pressure drop through the system. To find the total pressure drop through the total system the pressure drop through the supply register most distant from the blower and the pressure drop through the return air grille most distant from the blower must each be added to the duct pressure loss. The final result is the total pressure drop through the entire system.

For instance, if the equivalent length of the supply duct is 145 ft and the equivalent length of the return duct is 67 ft, sum of the two is found:

$$145 + 67 = 212$$

212 ft is the total equivalent length of the ductwork. This figure is then divided by 100:

$$212 \div 100 = 2.21$$

2.21 is the number of blocks of duct 100 ft long. If the ductwork was originally sized for 0.10 in. wg per 100 ft of duct, a common figure for small duct systems, then 2.21 is multiplied by 0.10 in. wg:

$$2.21 \times 0.10 = 0.212$$

0.212 in. wg is the friction drop in the duct.

To find the equivalent lengths of elbows, transitions, and other parts of a section of duct that is not straight the tables found in the Appendix in Table A-10 may be used.

To use the equivalent length tables, choose the fitting from the table that most nearly matches the fittings in the duct system. Then add the equivalent length given in the table for that fitting to the straight duct length. This calculation is made for each fitting in the system, and then the pressure drops through each of the fittings is added to the pressure drop through the duct sections. When the pressure drop through the supply register and return air grille are added to this figure the total pressure drop through the system is found.

16.6 SYSTEM INSTALLATION PROCEDURES

The designer should be involved in the installation of the heating system. Unless the system has been planned in great detail and coordination with all of the other building component designers has been possible, there will be a need for some flexibility in the installation of the ductwork and equipment. Quite commonly, beams and girders, water and sanitary piping, or electrical wiring will run through the same areas that the heating system designer has chosen for the ductwork. In such cases, the designer and the contractors installing the systems must work together to decide which system can best be changed so that both can be installed. Usually the designer or contractor can rearrange the work to accommodate the installation of both.

The installers of the systems must know exactly what the designer intends as far as performance is concerned. Slight changes in duct or pipe sizing can make a great difference in performance, and yet the installer may not appreciate that fact. If inspections are included in the design of the system, then the designer should make sure that the installer follows the original plan as far as possible.

16.7 SUMMARY

After the loads are figured and the tentative selection of equipment is made, the next step in designing a heating system is to select the distribution system to be used, lay out that system on a plan, size the ducts for air distribution, and check the blower capacity against the system requirements.

Three main types of air distribution systems are used: (1) perimeter loop, (2) radial, and (3) trunk and branch.

The perimeter loop system has a loop of duct that runs around the perimeter of a building, usually under the floor. This system is basically an under-floor system, and it is generally used in a concrete slab-on-grade construction. The main advantage of this system is that the loop duct warms the concrete slab around the outside perimeter of the building, making the concrete floor more comfortable.

In a radial duct system, branch ducts radiate out from a supply plenum, and each feeds a register. The branches usually are of small size since they each feed only one register. This system generally is installed in either a crawl space under a building, or an attic space.

In a trunk and branch duct system, the trunk is the main distribution member, taking the air off the supply plenum, and branches take air off the trunk to the registers. Two types of trunk and branch systems are used: extended plenum, and graduated trunk. The extended plenum system should be used only on small systems, while the graduated trunk system can be used on any size system.

Ducts should be carefully sized to ensure the proper delivery of air required into each part of a building. Duct sizing can be done by the use of an equal friction chart. Proper sizing also makes it easy to calculate the total resistance to airflow so the blower can be properly sized for the delivery of air.

When the heating system is being installed, the designer should be involved in the job to ensure that the components go in as planned. Sometimes changes have to be made to the system to facilitate the installation of other components of the building. If changes are made, they must not alter the basic function of the system.

16.8 TERMS TO KNOW

configuration
perimeter loop system
branch ducts
loop
radial system
trunk and branch system
radial duct system
extended plenum trunk and branch system
balance
graduated trunk system
branch take-off
static pressure
performance standards

CHAPTER
17
Warm Air Registers and Grilles

17.1 INTRODUCTION

In designing a warm air heating system, supply registers and return air grilles are selected, located, and sized to ensure that the correct amount of warm air is distributed to each space in the building, as determined by the heat load calculations. The required cfm for each room is used to select the registers and grilles. In this chapter the various supply registers and return air grilles available are described, and factors to consider when selecting, locating, and sizing the registers and grilles are covered.

17.2 SUPPLY REGISTERS

Supply registers are the components of the warm air heating system that introduce air into rooms or spaces to be heated. A supply register is made up of a **diffuser** or face, a frame to hold it, and a damper to control the quantity of air passing through it. A typical supply register is shown in Figure 17-1.

Register type, location in the room, and size are the most important factors to consider in the selection of the supply register for a system. The style of the register and its location in a space will determine how effectively the air is distributed. The size of the register will determine the amount of air that will be delivered.

Supply register*

Diffuser – A face grille, used on a register, with vanes that cause the air to spread as it leaves the face.

Ceiling register*

*Term defined in text.

Sidewall register*

Perimeter register*

FIGURE 17–1

Typical supply register used on a warm air system

Selecting Supply Registers by Location

There are three basic types of supply registers categorized by their location in a room:

1. Ceiling
2. Sidewall
3. Perimeter

Ceiling registers, as the name implies, are located on the ceiling of a room and are designed to deliver the air both horizontally and vertically. Figure 17-2 is an illustration of a typical square ceiling register.

Sidewall registers are located on a wall of a room, usually in one of two general locations: high on the wall near the ceiling or low near the floor. In a typical sidewall register, the air is diffused horizontally and vertically by the vanes in the grille. A sidewall register is shown in Figure 17-3.

Perimeter registers are located around the outside walls of the building, either on the floor or the ceiling. Perimeter registers deliver the air either vertically up or vertically down, depending on their location. A perimeter floor register is shown in Figure 17-4.

Ceiling Registers. Ceiling registers are used when the space to be heated is too large to heat adequately from the walls or along the walls only, or when the amount of air required is greater than can be provided from the wall locations. When registers are placed at the proper locations around the ceiling of a room, air can be circulated evenly regardless of the size of the room.

Ceiling registers can be either square or round in shape. Figure 17-5 shows a round ceiling register.

The ductwork for ceiling registers is located above the register. The shape of the register normally is determined by the construction of the ceiling or the shape of the room. Aesthetics, or looks, also are considered in the choice.

Although the air pattern from a ceiling register is basically vertically down, the face of the register is designed to spread the air so that it will cover a wider area. A diffuser type of register face accomplishes this purpose. Figure 17-6 shows the diffuser from a register. The curved vanes direct the airflow.

FIGURE 17–2
Ceiling register

FIGURE 17–3
Typical sidewall register

FIGURE 17–4
Floor register used in perimeter floor systems

FIGURE 17–5
Round ceiling register

FIGURE 17–6
Diffuser used on the face of a ceiling register to
deflect the air

Care must be taken in sizing and locating ceiling diffusers so the air from them will not strike the occupants of
the space in which they provide air, at too high a velocity,
or before the air has had time to mix with room air. Air of
either too high a velocity or too low a temperature will be
uncomfortable.

Sidewall Registers.   Sidewall registers are used when the
design of a building is such that registers can be located in
the walls, but the discharged air will not strike any occupant until it has mixed with room air. The registers are
located to deliver the air basically in a horizontal pattern,
but the register face is designed so the airstream is also
spread over a fairly large horizontal area. Figure 17-7 is an

FIGURE 17–7
Sidewall register

illustration of a sidewall register. The vanes are adjustable so that the desired air distribution pattern can be attained.

Location and sizing of sidewall grilles is critical to ensure comfort. Low sidewall registers should be used only when it is possible to direct the air so it does not enter the occupied space at too high a velocity for comfort.

Perimeter Registers.   The most commonly used register system for small commercial and residential systems is the perimeter floor or ceiling system. In this system, narrow slot-type diffusers are located in the floor or ceiling around the perimeter of the building so the air will cover the windows, doors, and any other area of high heat loss.

The air from perimeter registers is directed vertically up or down, but it also is spread by vanes in the face of the register. Since the registers are located around the walls of the room, the air usually will not strike any occupant of the space until it has had a chance to mix with the room air. The air distribution pattern, or spreading pattern, for a perimeter register is shown in Figure 17-8.

Selecting Supply Registers by Air Pattern

The style of supply register selected for any distribution system will be determined by the type of register that will deliver the air where it will best cover the area of heat loss, and provide the amount of air required. Four factors have to be considered:

1. cfm
2. Throw
3. Spread
4. Noise

Cfm, the air required in the room, has been discussed in previous chapters. **Throw** is the distance the air will travel directly away from the register before it reaches a terminal velocity of 50 fpm (ft per min). **Spread** is the width of the space that the air will cover, usually in a fan shape, before it reaches a terminal velocity of 50 fpm. **Noise level** of a register is related to the sound generated by the air passing through the vanes in the register face. Each of these factors has to be considered in the selection process.

Throw*

fpm – Foot per minute; a designation for velocity used in air distribution systems.

Spread*

Noise level*

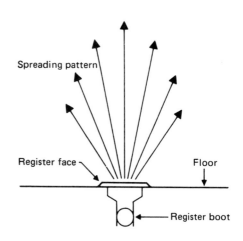

FIGURE 17–8
Spreading pattern of a perimeter register

cfm. The first consideration in selecting the supply registers is based on the fact that air is the medium for carrying the heat where it is wanted in the building. The proper cfm has to be delivered by each register in the system to ensure proper distribution of heat. The amount of air that any register is designed to deliver is shown in a register manufacturer's catalog. Table 17-1 is a page from a register catalog. The table shows how cfm is related to register size.

The amount of air that a register will deliver is determined by the opening size of the register. The velocity of air out of the opening is determined by the amount of air. Manufacturers' catalog selection sheets show the different types of registers with the cfm that each size will deliver and the velocity of the air out of the face. They also show the spread and throw, which will be discussed later. The following example shows how to use the register catalog data for register selection.

EXAMPLE 17-1

A room with the dimensions of 12 ft × 24 ft is found to need 250 cfm of air to heat it. It is decided that a ceiling diffuser will be used. What size register will deliver the required cfm?

Solution

Refer to the catalog page for ceiling diffusers (Table 17-1). The catalog page includes a register with a two-way pattern. The boldface numbers in the bottom section give the cfm for each register. The nearest cfm for this example is 252, provided by a 9 in. × 9 in. register. The register will deliver 126 cfm out of each side. The throw will be from 7 ft to 13 ft on each side. This register will deliver the wanted air in a way that will cover the room completely.

A designer may choose one or more registers for each area in a building, depending on the amount of air required. The choice of the number of registers is based on the amount of air and the air distribution pattern desired.

Throw. An important consideration in selecting registers, is how far and in what direction the air will travel after it leaves the register. The distance the air travels before its velocity becomes too slow to be effective is called throw. Figure 17-9 is an illustration of throw.

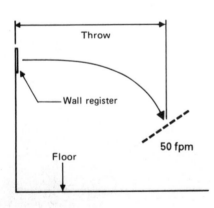

FIGURE 17-9
Air throw from a wall register

TABLE 17–1
Page from a Register Catalog

specifications—series 95   ONE WAY PATTERN

SIZE IN INCHES	VEL* PRESS†	400 / .044	450 / .056	500 / .068	550 / .083	600 / .100	650 / .114	700 / .130
CORE NO. 111								
6 x 6	TOTAL CFM	A 100 B	A 112 B	A 125 B	A 137 B	A 150 B	A 162 B	A 175 B
	CFM/side	100	112	125	137	150	162	175
	Throw, ft.	6-11	6-12	7-14	7-15	9-17	9-18	10-20
9 x 9	TOTAL CFM	224	252	280	308	336	364	392
	CFM/side	224	252	280	308	336	364	392
	Throw, ft.	7-13	7-15	9-17	9-18	10-20	11-22	12-24
12 x 12	TOTAL CFM	400	450	500	550	600	650	700
	CFM/side	400	450	500	550	600	650	700
	Throw, ft.	7-15	9-17	10-19	10-21	11-23	12-25	13-27
15 x 15	TOTAL CFM	624	702	780	858	936	1014	1092
	CFM/side	624	702	780	858	936	1014	1092
	Throw, ft.	9-17	10-19	10-21	11-23	13-26	14-28	15-30
CORE NO. 112								
6 x 9	TOTAL CFM	150	169	188	206	225	244	262
	CFM/side	150	169	188	206	225	244	262
	Throw, ft.	6-12	7-14	7-15	9-17	10-19	10-20	11-22
9 x 12	TOTAL CFM	300	338	375	413	450	488	525
	CFM/side	300	338	375	413	450	488	525
	Throw, ft.	7-14	8-16	9-18	10-20	10-21	11-23	12-25
9 x 15	TOTAL CFM	375	422	469	516	563	610	656
	CFM/side	375	422	469	516	563	610	656
	Throw, ft.	7-15	9-17	9-18	10-20	11-22	12-24	13-26
12 x 15	TOTAL CFM	500	562	625	688	750	813	875
	CFM/side	500	562	625	688	750	813	875
	Throw, ft.	8-16	9-18	10-20	11-22	12-24	13-26	14-28
12 x 18	TOTAL CFM	600	675	750	825	900	975	1050
	CFM/side	600	675	750	825	900	975	1050
	Throw, ft.	8-16	9-18	10-20	11-22	12-25	13-27	14-29
CORE NO. 113								
6 x 9	TOTAL CFM	150	169	188	206	225	244	262
	CFM/side	150	169	188	206	225	244	262
	Throw, ft.	6-12	7-14	7-15	9-17	10-19	10-20	11-22
6 x 12	TOTAL CFM	200	225	250	275	300	325	350
	CFM/side	200	225	250	275	300	325	350
	Throw, ft.	7-13	7-15	8-16	9-18	10-20	10-21	11-23
6 x 15	TOTAL CFM	250	281	312	344	375	406	438
	CFM/side	250	281	312	344	375	406	438
	Throw, ft.	7-13	7-15	9-17	10-19	10-20	11-22	12-24
6 x 18	TOTAL CFM	300	337	375	413	450	487	525
	CFM/side	300	337	375	413	450	487	525
	Throw, ft.	7-14	8-16	9-18	10-20	10-21	11-23	12-25
6 x 21	TOTAL CFM	350	394	437	481	525	570	612
	CFM/side	350	394	437	481	525	570	612
	Throw, ft.	7-14	8-16	9-18	10-20	11-22	12-24	13-26
6 x 24	TOTAL CFM	400	450	500	550	600	650	700
	CFM/side	400	450	500	550	600	650	700
	Throw, ft.	7-15	9-17	10-19	10-21	11-23	12-25	13-27
9 x 21	TOTAL CFM	524	590	655	720	786	852	917
	CFM/side	524	590	655	720	786	852	917
	Throw, ft.	8-16	9-18	10-20	11-22	12-24	13-26	14-28
9 x 24	TOTAL CFM	600	675	750	825	900	975	1050
	CFM/side	600	675	750	825	900	975	1050
	Throw, ft.	8-16	9-18	10-20	11-22	12-25	13-27	14-29

specifications—series 95   TWO WAY PATTERN

SIZE IN INCHES	VEL* PRESS†	400 / .044	450 / .056	500 / .068	550 / .083	600 / .100	650 / .114	700 / .130
CORE NO. 211								
6 x 6	TOTAL CFM	A 100 B	A 112 B	A 125 B	A 137 B	A 150 B	A 162 B	A 175 B
	CFM/side	50 50	56 56	62 62	69 69	75 75	81 81	88 88
	Throw, ft.	5-10 5-10	6-11 6-11	6-12 6-12	7-13 7-13	7-15 7-15	8-16 8-16	9-17 9-17
9 x 9	TOTAL CFM	224	252	280	308	336	364	392
	CFM/side	112 112	126 126	140 140	154 154	168 168	182 182	196 196
	Throw, ft.	6-11 6-11	7-13 7-13	7-14 7-14	8-16 8-16	9-17 9-17	10-19 10-19	10-20 10-20
12 x 12	TOTAL CFM	400	450	500	550	600	650	700
	CFM/side	200 200	225 225	250 250	275 275	300 300	325 325	350 350
	Throw, ft.	7-13 7-13	7-15 7-15	8-16 8-16	9-18 9-18	10-20 10-20	10-21 10-21	11-23 11-23
15 x 15	TOTAL CFM	624	702	780	858	936	1014	1092
	CFM/side	312 312	351 351	390 390	429 429	468 468	507 507	546 546
	Throw, ft.	7-14 7-14	8-16 8-16	9-18 9-18	10-20 10-20	10-21 10-21	11-23 11-23	12-25 12-25
18 x 18	TOTAL CFM	900	1012	1125	1240	1350	1462	1575
	CFM/side	450 450	506 506	562 562	620 620	675 675	731 731	788 788
	Throw, ft.	7-15 7-15	9-17 9-17	10-19 10-19	10-21 10-21	11-23 11-23	12-25 12-25	13-27 13-27

*Neck velocity fpm. Jet velocity will approximate neck velocity x 2.
†Total pressure (sum of velocity and static pressures) inches of water.

Reprinted with permission from Lima Register Co.

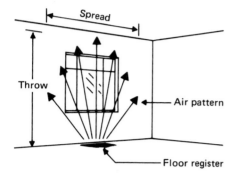

FIGURE 17-10
Air throw and spread from a floor register

The throw for a given register depends upon the velocity of the air at the register face. The velocity is a function of the quantity of air and the size of the register. The face area and the throw of the register are shown on the selection charts.

Some register catalogs show two face areas. One is the total area; it is based on the size of the opening in the register through which air can pass. The other area is the net face area; it is the total area less the area taken up by the vanes or any other obstructions in the face of the grille.

Registers must direct the air to where it is needed in the space. A terminal velocity of 50 fpm is considered the point at which throw is ended.

Spread. Most registers have a grille, or curved vanes, on the face to spread the air as it leaves the register. The amount of this spread is important in determining the air pattern that will be achieved. An illustration showing air throw and spread from a floor register is provided in Figure 17-10. Spread is the width of the fan of air that comes out of a register. The perimeter of the pattern is considered to be where the air has dropped to a velocity of 50 fpm or less.

Good design requires that the conditioned air cover the areas of greatest heat loss, such as windows, doors, and cold outside walls. Therefore, a designer should select the registers with the proper spread to cover these areas with a blanket of air.

Noise. Air passing through the face of any register go through the vanes in the grill face. These vanes are relatively small. If the quantity of air is too great, the vanes vibrate, like the strings on a harp. If the vibration becomes too great, noise is produced. To control the amount of noise, the designer must select registers large enough so the velocity of air will not create vibrations.

Most register manufacturers' selection charts are marked in such a way that they indicate how much air can be delivered by each size of register without excess noise. Table 17-2 is a copy of a register manufacturer's catalog sheet for floor registers. The chart is separated into two sections. All register selections for any given cfm on the left side of the double vertical lines will be quiet in operation; those on the right side may be noisy.

TABLE 17–2

Catalog Sheets for Registers Indicating Limits at Which They Will Operate Quietly

NOMINAL SIZE	FREE AREA SQ. IN.	Heating BTU/h	3045	4565	6090	7610	9515	11415	13320	15220	17125	19025	20930	22830	24735	26635	30440	34245	38050	45660
		Cooling BTU/h	855	1280	1710	2135	2670	3200	3735	4270	4805	5340	5870	6405	6940	7470	8540	9605	10675	12810
		C.F.M.	40	60	80	100	125	150	175	200	225	250	275	300	325	350	400	450	500	600
2¼″ x 10″	19	T.P. Loss	.013	.022	.035	.045	.060	.092	.120	.150										
		Vert. Throw (ft.)	3.0	4.5	5.5	6.5	8.5	10.5	13.0	15.5										
		Vert. Spread (ft.)	6	8	10	12	15	18	23	26										
		Face Velocity	309	463	617	771	964	1159	1352	1545										
2¼″ x 12″	21	T.P. Loss	.009	.015	.027	.037	.050	.080	.105	.134										
		Vert. Throw (ft.)	3	4	5	6	8	10	12	14										
		Vert. Spread (ft.)	6	8	10	11	14	17	22	25										
		Face Velocity	280	420	565	705	880	1050	1230	1400										
2¼″ x 14″	24	T.P. Loss	.006	.010	.021	.031	.042	.070	.093	.121	.150									
		Vert. Throw (ft.)	3	4	4.5	5.5	8	9.5	11	12.5	14									
		Vert. Spread (ft.)	6	8	9	11	14	16	19	22	25									
		Face Velocity	245	365	490	610	760	915	1065	1220	1370									
4″ x 10″	32	T.P. Loss		.008	.021	.026	.032	.045	.062	.084	.110	.134	.163							
		Vert. Throw (ft.)		3	4	5	7	8.5	10	11	12	13	14							
		Vert. Spread (ft.)		6	8	9	12	14	17	19	22	24	26							
		Face Velocity		265	355	445	555	665	775	890	1000	1120	1220							
4″ x 12″	39	T.P. Loss		.004	.010	.016	.023	.033	.042	.058	.075	.089	.107	.128	.159					
		Vert. Throw (ft.)		2	3	4	6.5	8	9	10	11	12	13	14	15					
		Vert. Spread (ft.)		4	6	8	12	14	16	18	20	23	24	26	28					
		Face Velocity		220	295	370	460	555	645	735	830	925	1020	1110	1200					
4″ x 14″	46	T.P. Loss			.006	.010	.016	.021	.028	.039	.051	.060	.080	.101	.124	.137	.167			
		Vert. Throw (ft.)			3	4	6	7	8	9	10	11	12	13	14	15	16			
		Vert. Spread (ft.)			6	8	11	12	14	16	18	20	22	24	26	28	30			
		Face Velocity			255	320	395	475	555	635	715	790	870	950	1025	1110	1270			
6″ x 10″	52	T.P. Loss	Series 40 Only			.009	.014	.019	.027	.035	.044	.054	.064	.078	.090	.104	.135	.171	.210	.314
		Vert. Throw (ft.)				4	5.5	6.5	7.5	8.5	10	11	12	13	14	15	17.5	19	21.5	26
		Vert. Spread (ft.)				8	9.5	11.5	13.5	15.5	17	19	21	23	25	27	31	34.5	38.5	46
		Face Velocity				278	348	417	487	556	626	695	765	834	904	973	1112	1251	1390	1668
6″ x 12″	59	T.P. Loss	Series 40 Only			.007	.011	.015	.021	.027	.035	.042	.050	.060	.071	.083	.107	.134	.165	.242
		Vert. Throw (ft.)				4	5	6	7	8	9	10.5	12.5	13.5	14	16.5	18.5	20.5		29
		Vert. Spread (ft.)				7	9	11	12.5	14.5	16	18	20	21.5	23.5	25	29	32.5	36	43
		Face Velocity				245	307	368	429	491	552	613	675	736	797	859	981	1103	1227	1472
6″ x 14″	66	T.P. Loss	Series 40 Only			.005	.009	.012	.017	.022	.028	.034	.041	.050	.058	.067	.088	.110	.132	.194
		Vert. Throw (ft.)				4	5	6	7	8	8.5	9.5	10.5	11.5	12	13	15	17	19	23
		Vert. Spread (ft.)				7	8.5	10	12	13.5	15	17	18.5	20.5	22	24	27	30.5	34	41
		Face Velocity				219	274	329	384	439	494	549	604	659	713	768	878	988	1098	1317

Reprinted with permission from Lima Register Co.

17.3 RETURN AIR GRILLES

A **return air grille** has two parts: a decorative grille or face and the frame that holds it and connects it to the ductwork. The face may have curved or louvered vanes to make it more attractive or, more likely, to obstruct the view through it. A typical return air grille is shown in Figure 17-11.

If a supply air system is properly designed and the supply registers are sized and located so they will do a

Return air grille*

FIGURE 17–11
Typical return air grille

good job of supplying the air to take care of the heat loss, then the function of the return air system is simply to provide a way for that air to get back to the air mover. The air mover, or blower, can deliver only as much air on the supply side as it takes in on the return side. Therefore, the return air grilles have to be located so that the air, as supplied by the system, can find its way to them.

The return air grilles should be located as near to the center of any building space which they serve, as possible. This placement keeps the friction in the return air duct as small as possible. If a return grille is located in the center of a zone, care should be taken to make sure that all of the air from the supply registers can return to it. If any of the supplies are located in rooms that may be cut off from the return grille location, then door grilles or wall grilles should be provided to allow that air to reach the location of the return grille.

Return air grilles are sized to handle the quantity of air in the zone they will serve at a low enough velocity to be quiet in operation. The acceptable velocity for a variety of applications is shown in Table 17-3.

When more than one return air grille is used for a zone, each grille should be sized to handle the amount of air required for its part of the zone at a face velocity below that shown on the chart. Door grilles, **transfer grilles** (wall grilles), or any other return air grilles should be sized in the same way. The return air system must be sized to carry at least the amount of air that the supply side will furnish. If the air handler cannot get air back from the building, it will not be able to supply it.

17.4 VENTILATION AIR AND EXHAUST GRILLES

Ventilation air and exhaust grilles, or louvers, should be sized in the same way as the other registers or grilles in a building. Each should be sized for the cfm required to flow through it and for a velocity of air that will be quiet and acceptable for each location.

Choosing the location of the ventilation air and exhaust grilles should be carefully done. Ventilation air grilles must be located so contaminated air from chimneys, flues, or vents will not be drawn into the building. Exhaust

Transfer grille – A grille installed in an opening between two rooms that allows air to pass between the rooms.

TABLE 17–3
Return Air Grille Selection Chart

Product Number and Size	Free Area Sq. In.	Maximum Recommended C.F.M.	Product Number and Size	Free Area Sq. In.	Maximum Recommended C.F.M.	Product Number and Size	Free Area Sq. In.	Maximum Recommended C.F.M.	Product Number and Size	Free Area Sq. In.	Maximum Recommended C.F.M.
18H- 6 x 4	16	65	18H-14 x 10	74	390	18H-18 x 18	216	900	18H-24 x 16	256	1060
18H- 8 x 4	21	89	18H-14 x 12	113	468	18H-20 x 6	77	318	18H-24 x 18	288	1200
18H- 8 x 6	32	135	18H-14 x 14	132	545	18H-20 x 8	104	432	18H-24 x 20	320	1330
18H- 8 x 8	43	180	18H-16 x 4	43	180	18H-20 x 10	130	540	18H-24 x 24	384	1600
18H-10 x 4	26	107	18H-16 x 6	65	270	18H-20 x 12	156	648	18H-30 x 4	78	324
18H-10 x 6	39	162	18H-16 x 8	87	360	18H-20 x 14	182	755	18H-30 x 6	118	486
18H-10 x 8	52	216	18H-16 x 10	109	450	18H-20 x 16	209	870	18H-30 x 8	158	654
18H-10 x 10	65	270	18H-16 x 12	130	540	18H-20 x 18	235	977	18H-30 x 10	199	828
18H-12 x 4	31	130	18H-16 x 14	152	630	18H-20 x 20	262	1090	18H-30 x 12	239	990
18H-12 x 6	47	192	18H-16 x 16	174	725	18H-24 x 4	63	258	18H-30 x 14	279	1159
18H-12 x 8	63	258	18H-18 x 6	71	294	18H-24 x 6	95	390	18H-30 x 16	319	1325
18H-12 x 10	79	324	18H-18 x 8	95	396	18H-24 x 8	127	528	18H-30 x 18	360	1500
18H-12 x 12	96	396	18H-18 x 10	119	491	18H-24 x 10	159	660	18H-30 x 20	399	1660
18H-14 x 4	37	150	18H-18 x 12	144	600	18H-24 x 12	191	792	18H-30 x 24	479	1990
18H-14 x 6	56	228	18H-18 x 14	167	696	18H-24 x 14	224	930	18H-30 x 30	600	2500
18H-14 x 8	75	312	18H-18 x 16	192	797						

Reprinted with permission from Lima Register Co.

grilles must be located so exhaust air will not recycle into the ventilation air grilles, or be blown into any occupied space.

17.5 REGISTER PLANS

Layouts of the distribution and return air system should show the register sizes. The type of register to be used at each location, the cfm to be delivered by the register, and the size of the register should be lettered in by the register symbol or indicated on a separate schedule. These data will be used in figuring the cost of the job, in laying it out, and also in balancing the system for proper airflow.

17.6 SUMMARY

The supply registers in a warm air heating system introduce the heated air, and provide the proper distribution of heated air into the space to be heated. The selection, location, and size of the registers is of paramount importance in the design of a good heating system. The return air grilles are used to collect the air from the building and return it to the blower so a continuous supply of air can be provided.

Supply registers come in three main types: ceiling, wall, and perimeter. Ceiling registers are used when having the air distributed through the center area of the room is desired. Wall registers are used when the air can be distributed from one or more sides of a room and still cover the entire space. Perimeter registers are used to blanket the outer surfaces of a building with warm air to offset the greater heat losses that occur there.

Registers also are designed to spread the air over a wide area. The throw and spread of that area is considered when a selection is made of a register for a particular application.

Supply registers are sized to supply the amount of air required at a given location at a high enough velocity to direct it where it is needed. The selection should also be made so that the velocity will not be so great that air noise at the register face becomes a problem.

Return air grilles are located in a building so they return all of the air that the supply side distributes. A supply blower will put out only as much air as it can get back from the return side of the system. The return air grilles must be located so that air from each room or space with a supply register can get to them. They are sized for quantity of air and velocity at the face. The face velocity must be low enough so the grilles are quiet.

Ventilation air or exhaust air grilles, or louvers, are located and sized in the same way as supply registers and return air grilles.

17.7 TERMS TO KNOW

supply registers
diffuser
ceiling register
sidewall register
perimeter register
throw
fpm
spread
noise level
return air grille
transfer grille

18.1 INTRODUCTION

An installing contractor often is unable to install a warm air heating system in exactly the way that the designer draws it up on the mechanical plan. Unforeseen complications, such as changes in construction or conflicts with plumbing or electric systems, can require the ductwork to be changed. When such changes are made in the system, there will be more or less friction than was originally planned. Because the duct has been sized to provide a specific amount of air into each room or zone based on the predicted friction in the system, these changes will alter the actual amount of air flowing. Adjustments must be made to the system to compensate for these changes. Also, since the blower and the registers are not factor adjusted for a specific amount of air, they must be adjusted on the job.

An air balance for a warm air heating system includes adjusting the blower so it puts out the total air required for the system, making sure all branches in the system have the proper amount of air flowing in them, and adjusting each register so that it has the amount of air flowing out of it that was calculated when the heat load was converted to cfm. Each of these operations is explained and described in this chapter.

18.2 TERMS RELATED TO AIRFLOW

Some specific terms related to airflow, blowers, and duct systems are important to an understanding of airflow and measurement of air in a duct system.

Friction

As air flows through a duct, there is always friction between the walls of the duct and the film of air in contact with them. See Figure 18-1.

There also is resistance to the airflow in any fitting in the system that requires the air to change speed or direction, such as **take-offs, reducers,** or **elbows.** All duct resistance to airflow can be termed **friction.** Friction can be measured by the effect on the static pressure in the system.

CHAPTER

18
Air Balance

Take-off – A duct fitting designed to allow air to flow from a trunk to a branch.

Reducer – A section of air duct where the size of the duct becomes smaller.

Friction*

*Term defined in text.

FIGURE 18–1 _____
Resistance caused by friction between the air flowing in a duct and the walls of the duct

Static Pressure

Static pressure (sp) describes the amount of air pressure in a duct as measured at right angles to the flow of the air. It can be defined as the pressure that would have a tendency to burst the duct. Figure 18-2 illustrates how static pressure is measured.

Static Pressure Drop

The difference between the static pressure at two different points in a duct, or between the static pressure in the duct and the pressure in a room, is called the static pressure drop. The resistance in a duct causes a static pressure drop to occur as the air passes along the duct (disregarding changes in velocity), so static pressure drop can be used to measure friction. Figure 18-3 demonstrates how static pressure drop is measured.

Static pressure drop can be represented by the following formula:

$$SPD = P_1 - P_2$$

where:

SPD = static pressure drop
P_1 = static pressure at one point in a duct
P_2 = the static pressure at a point downstream from P_1

EXAMPLE 18-1

The static pressure read at the blower discharge on a typical duct system is found to be 0.06 in. wg. This pressure is compared to the pressure at the discharge of the duct, which will be 0.00 in. wg (room pressure). Find the static pressure drop in that section of duct.

Solution

Use the following formula:

$$SPD = P_1 - P_2$$

FIGURE 18–2 _____
Measuring static pressure

FIGURE 18–3 _____
Measuring static pressure drop

FIGURE 18–4 _____
U-tube manometer

where:

    SPD = static pressure drop in the duct
    P_1 = pressure at blower discharge (in this example, 0.06 in. wg)
    P_2 = pressure in the room at duct discharge (in this example, 0.00 in. wg)

Therefore;

    SPD = 0.06 - 0.00 = 0.06

The static pressure drop in the duct is 0.06 in. wg. This measurement indicates the amount of friction in the duct.

Manometer or Water Gauge

The static pressure in a duct is measured with a manometer, a device that measures air pressure. Basically, a manometer is a U-shaped tube containing water or some other liquid. If a fluid other than water is used, the scale is calibrated to give a reading corresponding to one for water. See Figure 18-4 for a diagram representing a manometer.

Air pressure

Pressure in inches of water

4"
3"
2"
1"
0
1"
2"
3"
4"

FIGURE 18–5
Using a manometer to measure air pressure

When a manometer is used, one end of a rubber tube is connected to one of the upright legs of the U. The other end of the tube is connected to hole in the side of the duct. Pressure exerted by the air against the inside wall of the duct is transferred through the tube, and causes the water in the one leg of the U to be depressed. A corresponding rise will be observed in the water in the other leg.

The difference between the level of the water in the two legs of the U is an indication of the pressure exerted by the air in the duct. The pressure is measured in inches and so is called the static pressure in inches of water (in. wg). See Figure 18-5.

Measuring Static Pressure to Determine Friction

To measure the static pressure drop in a duct, the water gauge pressure is read at each end of the duct. The difference between the two readings indicates the amount of friction in the duct. The pressure reading in the upstream part of the system must be higher than in the downstream part or the air would not move. The pressure is read by placing the tube at right angles to the flow of air and flush with the inside of the duct so that only the pressure exerted against the side of the duct is read.

18.3 CHECKING THE QUANTITY OF AIR IN THE SYSTEM

The first step in performing an air balance is to check the cfm output of the blower and adjust the output as necessary to deliver the amount of air required for the system. This operation should always be done with all balancing dampers and register dampers in the system set wide open.

When selecting the heating unit for a job, the system designer chooses a unit with a blower with the capacity to provide the amount of air needed for the job. Blower characteristics are such that any given blower will deliver air over quite a wide range of cfm and for different static pressure drops. When starting an air balance, the installer must determine the actual cfm output for the conditions on the particular system being balanced.

The air moving capacity of the blower used in a typical low pressure system is a function of the size of the blower wheel and of the speed of rotation of the wheel. The blowers most often used for heating systems are double-inlet, forward-curved type. Double inlet means that air will enter the blower from both sides. Forward curved relates to the angle at which the blades of the wheel are inclined to the direction of the air discharge. A typical blower is shown in Figure 18-6.

FIGURE 18–6
Typical forward-curved, belt-drive blower

Manufacturers provide data sheets with their equipment that include blower performance charts. These charts give the output of the blowers in cfm, for different static pressure drops, and blower revolutions per minute (rpm). These data are used to make the original selection of blower for the job. A copy of one such blower table is shown in Table 18-1.

There are two common methods of determining the cfm output of the blower. One is to measure the temperature rise of the air through the unit and then use the sensible heat formula to find the cfm mathematically. The other is to find the rpm of the blower, measure the static pressure drop in the duct system, and then find the cfm from the blower table. Both of these methods are explained in the following sections.

Using the Temperature Rise Method

To determine the cfm output of a blower with the temperature rise method, find the Btu/h output of the unit by reference to the manufacturer's data (the output can be double checked by finding the fuel use rate and calculating the actual output). Then find the temperature rise through the unit by taking the dry bulb temperature of the air entering the furnace and the dry bulb temperature of the air leaving the furnace, find the difference between the two, and use the sensible heat formula, as follows:

$$\text{cfm} = \frac{\text{Btu/h}}{(T_0 - T_1) \times 1.08}$$

TABLE 18–1
Blower Data from a Furnace Manufacturer's Catalog

DRIVE KITS

Usage	Drive Kit Model No.	Motor hp	Motor Pulley (in.) & Groove	**Blower Pulley (in.) & Groove	*Rpm Range	Belt	Shipping Weight (lbs.) 1 Package
Heating	DK 1/3-5 (BM-5790)	1/3	1/2 x 3-3/4 — OA	1 x 7 — A	590 — 835	4L410	17
3-1/2 & 4 tons	DK 1/2-6 (BM-5791)	1/2	5/8 x 4-1/8 — OA	1 x 7 — A	690 — 935	4L420	22
5 tons	DK 3/4-7 (BM-5792)	3/4	5/8 x 4-1/8 — OA	1 x 7 — A	690 — 935	4L420	27
	DK-2007 (BM-7523)	†1	5/8 x 4-3/4 — OA	1 x 7 — A	840 — 1085	4L430	38

*At 1725 rpm motor speed.   †Fan relay included in kit.
**Factory installed in furnace package and not included in drive kit.

G11E-200V BLOWER PERFORMANCE

STATIC PRESSURE EXTERNAL TO UNIT (Inches Water Gauge)

Air Volume (Cfm)	0		.10		.20		.30		.40		.50		.60		.70		.80		.90		1.00	
	RPM	BHP	RPM	BHP	RPM	BHP	RPM	BHP	RPM	BHP	RPM	BHP	RPM	BHP	RPM	BHP	RPM	BHP	RPM	BHP	RPM	BHP
1400	460	.23	530	.29	600	.35	665	.41	725	.46	775	.51	830	.57	880	.66	920	.72	965	.79	1005	.86
1600	520	.31	590	.38	650	.43	715	.49	765	.56	815	.62	865	.71	910	.79	955	.86	995	.93	1035	1.0
1800	580	.40	650	.46	705	.53	760	.61	810	.69	860	.78	905	.86	950	.94	990	1.02	1030	1.09	---	---
2000	650	.51	705	.58	760	.67	815	.77	860	.86	905	.95	950	1.03	990	1.10	---	---	---	---	---	---
2200	715	.65	765	.74	820	.86	870	.96	910	1.04	950	1.12	---	---	---	---	---	---	---	---	---	---
2400	780	.85	825	.95	875	1.07	915	1.14	---	---	---	---	---	---	---	---	---	---	---	---	---	---
2600	840	1.02	885	1.13	---	---	---	---	---	---	---	---	---	---	---	---	---	---	---	---	---	---

NOTE — All cfm is measured external to the furnace with the air filter in place.

Reprinted with permission from Lennox Industries Inc.

where:

cfm   = quantity of air in cu ft per min
Btu/h = heat content of the air in Btu per hr
T_0 = temperature of the air at the outlet of the
         furnace
T_1 = temperature of the air at the inlet of the furnace
1.08   = constant that includes the sensible heat of
         air, the specific weight, and 60 minutes in
         an hour

EXAMPLE 18-2

If the Btu/h output of a furnace is found to be 200,000
Btu/h, the return air temperature is 76°F and the supply
plenum air temperature is 154°F, find the cfm.

Solution

$$cfm = \frac{Btu/h}{(T_0 - T_1) \times 1.08}$$

$$= \frac{200,000}{(154 - 76) \times 1.08}$$

$$= 2374$$

The airflow through the furnace is 2374 cfm.

Using the Blower rpm Method

The second method of checking the cfm output of a heating unit is by determining the blower rpm, measuring the
static pressure in the system, and then using the manufacturers' blower tables to find the cfm. The rpm of the
blower is most easily found by using a **tachometer.** This
device gives the rpm of the motor directly. If a tachometer is not available, the rpm can be determined for
direct-drive motors by reading the rpm data on the motor
nameplate. If the blower is a belt-drive blower, the rpm is
a function of the motor speed and the motor and blower
pulley sizes. Figure 18-7 shows a furnace with a
belt-drive blower.

   To find the rpm of the blower from the motor rpm and
pulley sizes, use the following formula:

FIGURE 18–7
Furnace with a belt-drive blower

$$rpm_b = \frac{rpm_m \times pd_m}{pd_b}$$

where:

rpm_b = rpm of the blower
rpm_m = rpm of the motor
pd_m = diameter of the motor pulley
pd_b = diameter of the blower pulley

The motor rpm can be found by reference to the motor nameplate, and the pulley sizes can be measured. The diameters used should be the effective diameter across the belt where it fits on the pulley.

EXAMPLE 18-3

In a particular application, the motor speed on a belt-drive blower is determined to be 1725 rpm, the pulley diameter on the motor is 3-1/2 in., and the blower pulley diameter is 6 in. Find the blower rpm.

Solution

$$rpm_b = \frac{rpm_m \times pd_m}{pd_b}$$

$$= \frac{1725 \times 3.5}{6}$$

$$= 1006.2$$

The rpm of the blower in this application is 1006.2.

After the rpm of the blower is determined, the total static pressure drop in the system is found by reading the static pressure at the blower inlet and also at the blower outlet. The pressure reading at the inlet will be a negative number since this is the static pressure drop in the return air side of the system. The reading taken on the outlet side will be positive since this is on the supply side of the system. To find the total static pressure drop, these two figures should be added together mathematically *without regard to the signs*. The total is the total static pressure drop in the system. Reference can then be made to the manufacturers' blower charts, and the cfm can be·found by comparing the rpm of the blower to the static pressure drop in the system.

EXAMPLE 18-4

Find the static pressure drop for a blower with an rpm of 1006.2. The static pressure in the supply side of a system is 0.25 in. wg by measurement. The pressure in the return side is -0.15 in. wg.

Solution

Add the two water pressures together without regard to the signs:

$$0.25 + 0.15 = 0.40$$

The total static pressure drop is 0.40 in. wg. This static pressure is used with the rpm to find the cfm from the blower performance chart.

18.4 ADJUSTING THE BLOWER FOR PROPER cfm

If, when the blower output is checked, the total cfm is not correct for the system, adjustments must be made to get the correct amount. The blower speed will need to be changed.

The characteristics of the blowers used in most heating units are such that the quantity of air moved, in cfm, is directly related to the rpm of the blower wheel. To increase the volume of air being delivered, the speed of rotation of the wheel is increased. To decrease the volume of air, the speed is decreased. Because the volume of air is directly related to the speed of rotation, the relationship can be shown mathematically, as follows:

$$\frac{cfm_1}{cfm_2} = \frac{rpm_1}{rpm_2}$$

FIGURE 18–8
Furnace with a direct-drive blower

where:

cfm_1 = original blower output
cfm_2 = new blower output
rpm_1 = original blower speed
rpm_2 = new blower speed

The desired blower speed for the corrected cfm can be taken directly from the manufacturer's blower performance chart. The blower rpm is then set at this speed.

Changing cfm on a Direct-Drive Blower

If a blower is direct-drive, that is, the blower wheel is mounted directly in the motor shaft as seen in Figure 18-8, the blower speed is the same as the motor speed. The speed of the motor can be changed by rearranging the electrical leads to the motor. The speeds related to the different electrical connections are shown in the manufacturer's data sheets and usually on the motor nameplate. Since the speeds are constant at each different step, the closest speed to give the desired cfm must be used.

Changing cfm on a Belt-Drive Blower

To change the cfm on a unit with a belt-drive blower, the rpm of the blower is changed by changing pulley sizes. The blower pulley diameter is usually two to three times the diameter of the motor pulley. Figure 18-9 shows a typical pulley-belt arrangement. Because of the size difference, the blower turns more slowly than does the motor. When the speed of the blower must be changed, the size of the motor pulley is changed.

Sometimes adjustable motor pulleys are used. When this is the case, the speed of the blower can be changed by changing the diameter of the adjustable pulley. To make this change, a setscrew on the pulley is loosened, and the outer part of the pulley is turned in or out on the threaded hub. See Figure 18-10. Turning the outer part of the pulley moves it in closer to the inner part or moves it farther away. The effective diameter of the pulley is changed by allowing the belt to ride closer in the hub or causing it to ride farther out. When more adjustment is needed than the adjustable pulley will give, a motor pulley of a different size must be chosen.

Motor pulley   Belt   Blower pulley

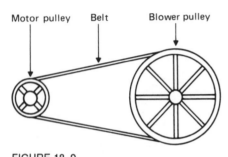

FIGURE 18–9
Typical pulley-belt arrangement

The speed of rotation of the blower wheel is directly related to the speed of rotation of the motor that drives it, but since the two are connected by the belt that runs between the two pulleys, the speed of the blower is indirectly proportional to the diameter of the two pulleys. Mathematically stated, this is

$$\frac{rpm_m}{rpm_b} = \frac{pd_b}{pd_m}$$

where:

rpm_m = revolutions per minute of the motor
rpm_b = revolutions per minute of the blower
pd_b = diameter of the blower pulley
pd_m = diameter of the motor pulley

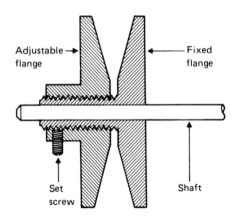

FIGURE 18–10
Adjustable pulley used for changing the speed of rotation of a blower wheel

This formula is the same as that used to find the original speed of the blower. Here, the elements are transposed to show the rpm as a ratio of the inverted pulley diameters.

Since the new rpm has already been determined from the blower performance chart, the only calculation necessary is to find the new motor pulley size for the required rpm. To do this operation, the formula is transposed to solve for pd_m, as follows:

$$pd_m = \frac{pd_b\ (rpm_b)}{(rpm_m)}$$

where

pd_m = diameter of the motor pulley
pd_b = diameter of the blower pulley
rpm_b = revolutions per minute of the blower
rpm_m = revolutions per minute of the motor

EXAMPLE 18-5

In Example 18-3, the following data were determined:

pd_m = 3.5
pd_b = 6
rpm_m = 1725
rpm_b = 1006.2

Use the data as a starting point. If a new blower speed of 1260 rpm was required, as found from a blower performance chart, find the proper motor pulley to give this speed.

Solution

$$pd_m = \frac{pd_b \, (rpm_b)}{(rpm_m)}$$

$$= \frac{6(1260)}{(1725)}$$

$$= 4.38$$

The motor pulley diameter should be changed to 4.38 in. to provide 1260 rpm, which will then increase the cfm to that required.

This change also will make a change in the static pressure drop in the system. If the change is not too great, then the blower performance chart will still be accurate enough for proper operation of the system.

After the motor pulley size is changed, the cfm should be checked again by the temperature rise method to make sure the cfm is acceptable.

18.5 BALANCING SYSTEM BRANCHES

Most small duct system layouts are quite simple. The system has a main trunk and individual branches to each register. The individual branches are sized to carry enough air to supply only one register. Balancing is done at the register damper. In larger systems and in better quality small systems, balancing dampers are installed in the branches.

If the branch serves only one register, then the balancing damper is used in the same way as a register damper. Many larger systems have multiple register takeoffs from each branch; many such branches may take off from the trunk. In such systems, the airflow in each branch must be regulated if the entire system is to remain in balance. The balancing dampers are found in the branch ducts just after the takeoff from the trunk.

Measuring Airflow in Branches

All of the balancing dampers in the branches should have been open when the blower was adjusted for design airflow. It is now necessary to take each branch duct in turn and measure the airflow in it and then adjust the dampers for the correct flow according to the design. Various instruments are used in this operation.

Measuring Instruments.   The most common method of measuring airflow in a duct is by use of a manometer and a special tube, called a **Pitot tube,** that is used to measure the velocity pressure of the air in the duct. The manometer used is the same as described earlier in this chapter, only the bottom part of the U is elongated and sloped. This shape allows the vertical height reading of the water to be spread over a longer distance, and the scale for reading it is larger and more accurate. This type of a manometer is called an **inclined manometer.** See Figure 18-11.

The special tube used with the manometer for measuring velocity is the Pitot tube. The Pitot tube gives pressure reading related to the velocity of the air in the duct. The reading obtained is called the **velocity pressure** (P_v) reading. We have already discussed static pressure (P_s). Static pressure plus velocity pressure equals **total pressure** (P_T). Velocity pressure is that pressure exerted by the air in a direct line with the airflow. Because of the nature of air pressure, velocity pressure cannot be measured directly. But it is possible to measure static pressure and total pressure. Since velocity pressure is the difference between these two pressures, it can be found from them by the use of the following formula:

$$P_v = P_T - P_s$$

where:

P_v = velocity pressure
P_T = total pressure
P_s = static pressure

A Pitot tube performs this operation automatically.

A Pitot tube is constructed so that one tube is inside another. The inner tube extends to the end that is aimed

Pitot tube – A tube used in measuring airflow that gives a total pressure reading and a static pressure reading simultaneously.

Inclined manometer*

Velocity pressure – The pressure exerted by a moving column of air in a direct line of flow.

Total pressure – The pressure exerted by a moving column of air in a direct line of flow. (Total pressure includes both velocity pressure and static pressure.)

FIGURE 18–11
Inclined manometer

Static pressure

Static pressure

Total pressure

Total pressure

FIGURE 18–12 _____
Diagram of a Pitot tube

Static pressure   Pitot tube

Total pressure

Velocity pressure reading ———

Manometer ———

FIGURE 18–13 _____
Tubes of the Pitot tube connected to the
manometer in order to get velocity pressure
reading

into the airstream and is open on the end. See Figure 18-12 for a diagram of a Pitot tube. The diagram shows the construction, location of the pressure taps, and points where each pressure is read. This tube measures total pressure because it reads both velocity pressure and static pressure simultaneously. The outer tube is closed at the end but has several holes around its outside perimeter so that it reads only static pressure.

The two tubes from the Pitot tube are connected to the manometer. See Figure 18-13. When they are connected in this way, the pressure exerted in the tubes opposes each other. The result is the velocity pressure reading, shown directly on the manometer scale. The velocity pressure can then be used to find the actual velocity by formula.

Calculating the cfm. Three steps are required to find the quantity of air flowing in a duct. First, the cross-sectional area of the duct must be found in sq ft. Second, velocity pressure is converted to velocity in fpm. Third, the area and the velocity are used to find the cfm.

In the first step, the area of the duct is found by multiplying the width of the duct in in. times the height in in., and dividing the results by 144 inches (number of in. in 1 sq ft):

$$A = \frac{w \times h}{144}$$

The second step is to use the following formula to find the velocity of air in the duct from the velocity pressure:

$$fpm = 4005 \sqrt{P_v}$$

where:

    fpm = velocity of the air in ft per min
    4005 = constant for standard air
    P_v = velocity pressure from manometer reading

The third step is to convert the velocity to cfm using this formula:

$$Q = Av$$

where:

Q = quantity of air in cfm
A = area of the duct in sq ft
v = velocity of the air in fpm

EXAMPLE 18-6

A 24 in. × 10 in. duct provides air to a small zone in a building. Upon measurement by manometer and Pitot tube, the velocity pressure (P_v) is found to be 0.048 in. wg. Determine the cfm.

Solution

First, find the cross-sectional area of the duct:

$$A = \frac{w \times h}{144}$$

$$\frac{24 \times 10}{144} = 1.67$$

The cross sectional area of the duct is 1.67 sq. ft.

Second, find the velocity of the air in the duct:

$$fpm = 4005 \sqrt{P_v}$$

$$fpm = 4005 \sqrt{0.048}$$

$$= 877.5$$

The velocity of air in the duct is 877.5 fpm.

Third, find the cfm (The area (A) of the duct must be in sq ft):

$$Q = Av$$
$$= 1.67 \times 877.5$$
$$= 1465$$

The quantity of air actually flowing in the duct is 1465 cfm.

Adjusting the cfm

If the existing cfm calculated for a branch is different from the design requirements, the adjusting damper needs to be changed to provide the correct amount. Finding the proper position may be a matter of some experimentation, but by checking the velocity pressure as the damper is changed, the correct cfm can be obtained.

EXAMPLE 18-7

Assume in Example 18-6 that the required cfm is 1330. The 1465 found in the example would be too much air. The damper should be closed somewhat to reduce the flow.

Solution

To simplify this procedure, first find the desired P_v for the 1330 cfm using the formula:

$$P_{v1} = \frac{P_{v2}\,(cfm_1)}{(cfm_2)}$$

where:

P_{v1} = velocity pressure desired for new cfm
P_{v2} = velocity pressure for existing cfm
cfm_1 = desired cfm
cfm_2 = existing cfm

Therefore;

$$P_{v1} = \frac{0.048\,(1330)}{(1465)}$$

$$= 0.044$$

The desired velocity pressure is 0.044 in. wg. The damper should be regulated until that reading is found on the manometer.

18.6 BALANCING THE SUPPLY REGISTERS

After the blower has been adjusted so that the correct amount of air is provided to the system and the branch adjustments are made, the next adjustment is that of the individual supply registers. Each register should be set so that it distributes the proper amount of air to the room or space that it serves. This balancing is done by adjusting the balancing dampers in the branch ducts or in the registers themselves.

Anemometer*

Velometer*

The airflow out of each supply register can be measured with any one of several instruments especially designed for this purpose. One is a special fitting that is used with a manometer. The fitting goes over the register face, and the manometer gives the velocity of the air in fpm. A second type is an **anemometer.** This device measures airflow by measuring the effect of the air on an electronic sensing device. Another type is a **velometer,** which reads the velocity of the air directly.

Instruments for Measuring Air at Supply Register Faces

When a manometer is used for measuring air quantity, a special fitting is used on one of the tubes. This fitting is placed over the register face so that the air passes through it. The fitting converts velocity pressure of the air from the register to static pressure. The manometer is calibrated to read in fpm.

An anemometer is an electronic device in which an electric current flows through a thermistor or similar sensing element. This sensing element is held in the airstream. The flow of air across the sensor affects the electric current. The effect is measured by a meter, calibrated to give a reading in fpm or cfm. Figure 18-14 shows a technician reading cfm from a register with an anemometer.

A velometer has a propeller rotated by a flow of air or a vane deflected by the flow. The propeller or vane is connected to a needle through a series of connecting links. The needle shows the velocity of the air on a calibrated scale. The velometer gives the velocity in fpm. Figure 18-15 shows a typical velometer.

One other meter used for checking cfm out of a register is shown in Figure 18-16. This device is a deflecting vane type of meter, but a scale on the indicating dial makes it possible to dial in the register size and read the cfm directly. The meter is used by placing it in the front of the register, and the air from the register flow through the meter.

Finding cfm

After the velocity of the air at a register face has been found, the quantity of air is calculated by using the velocity and face area of the register in the following formula:

FIGURE 18-14 _____
Using an anemometer to determine cfm from a register

FIGURE 18-15 _____
Deflecting vane velometer used to find fpm at a register face or in any airstream

FIGURE 18–16
Velometer used over the face of a register to find the cfm flowing in a direct reading

$$Q = Av$$

where

Q = quantity of air in cfm
A = area of the net face opening in sq ft
v = velocity of air in fpm

The free area of the register opening is given in the manufacturer's catalog. The balancer should have access to the proper performance chart for each type of register that is balanced.

EXAMPLE 18-8

Find the cfm through a 2 in. × 14 in. floor register with a velocity found in the manufacturer's catalog to be 760 fpm. The face area of the register is 23.6 sq in.

Solution

The area in sq in. needs to be changed to sq ft:

$$A = \frac{sq\ in.}{144}$$

$$= \frac{23.6}{144}$$

$$= 0.1639$$

The face area of the register in sq ft is 0.1639. This figure, with the velocity 760 fpm, is used in the velocity formula:

$$Q = Av$$
$$= 0.1639 \times 760$$
$$= 124.56$$

The volume of air out of the register is 124.56 cfm.

18.7 SUMMARY

After an air distribution system has been designed and installed, that system must be balanced for proper operation. An air balance is the process of adjusting the various air handling components in the system to deliver the desired quantity of air to each room or space in the building.

The three main components of the system that need checking and adjusting are the blower, the branch ducts, and the registers.

To start an air balance, all dampers in the system should be opened to ensure unrestricted flow of air. Then the blower is adjusted so it will provide the amount of air required, as determined when the system was designed. Next, the airflow is adjusted in each of the branch ducts in the system using the balancing dampers in the ducts. Third, the airflow out of each register is adjusted to match that required in the original design.

18.8 TERMS TO KNOW

take-off
reducer
friction

static pressure
tachometer
pitot tube
inclined manometer
velocity pressure
total pressure
anemometer
velometer

19.1 INTRODUCTION

A hydronic heating system is one in which heating is accomplished by the use of liquids. In most cases, the heat is generated at a central location in a building, and water carries the heat for use in other locations in the building. A piping system carries the water. At the points of use, air is heated by the hot water and then used for heating.

The many similarities between the design of the warm air heating system and the hydronic heating system will become apparent as the hydronic system is described. In this chapter, the major components of a hydronic system are described and their function in the system explained.

19.2 HEATING LOADS

The first step in designing any heating system is to find out how much heat must be furnished to offset the heat loss of the building. The heat load is found by accurately calculating the losses, as already described in Chapters 13 and 14.

Each part of a hydronic system is selected and sized to provide the correct amount of heat to offset the heat loss. The calculations must be made carefully and accurately. For design purposes, the heat load should be calculated for individual rooms or zones in the building, not for the building as a whole.

19.3 MAJOR COMPONENTS

The major components of a hydronic heating system are a boiler, distribution piping, circulating pumps, terminal devices, and two control systems. See Figure 19-1 for a diagram showing these components in a typical system. Each of these components is described in this chapter. The boiler and the controls are covered in detail in Chapter 10, however, and so are only mentioned briefly here.

19.4 BOILER

The boiler is the part of a hydronic heating system in which hot water is generated for the heating of the spaces

CHAPTER
19
Hydronic System Components

FIGURE 19–1 _____
Major components of a hydronic heating system

in the building. The construction and performance of the major types of boilers are covered in Chapter 10. Readers may want to review that description.

19.5 DISTRIBUTION SYSTEM

The second major component in the system is the distribution, or piping, system. The **distribution system** is the part of the system in which the hot water is transported from the boiler to the various points of use.

Several basic piping layouts can be used for hydronic distribution systems. The major types are

1. One-pipe, series loop system
2. Primary-secondary system
3. Two-pipe, direct return system

Distribution system*

*Term defined in text.

4. Two-pipe, reverse return system
5. Three- and four-pipe system.

Each of these is described in the sections that follow.

One-Pipe, Series Loop System

A **one-pipe, series loop system** is one in which a single pipe runs entirely around the building or a zone within a building. The terminal units are piped in series so that all of the water goes through each unit. Figure 19-2 illustrates a one-pipe, series loop system.

This system is simple and inexpensive compared to other systems. There are two major objections to its use: (1) All of the water for the entire system flows through each terminal device. Therefore, each device must be sized large enough to carry the total flow. (2) The temperature drop (TD) of the water through each unit affects all other units from that point on in the system. This system is practical only for the smallest jobs where the TD through the entire system is not too great.

Primary-Secondary System

A **primary-secondary system** is similar to the series system in that a single pipe carries the heating water around the building or zone, but the heating units are not in series on the loop. They are in parallel with it. An example of a typical primary-secondary system is shown in Figure 19-3.

In a primary-secondary system, each terminal device is connected to the main pipe with a supply and return connection. Although the TD of each individual unit affects the water temperature in the main line, the entire water supply does not flow through each unit. The main pipe is sized for the total gpm for the building or zone, but the secondary piping to each unit is sized to carry the gpm for that unit only.

The use of this system allows smaller piping for each unit and also makes it easier to provide proper temperature control in each separate zone served by the units.

The one-pipe, primary-secondary system is best used in small- to medium-sized buildings where more flexibility of control is needed than can be attained with a series loop. If the building is too large to heat with one loop, then

FIGURE 19–2
One-pipe, series loop distribution system

FIGURE 19–3
Primary-secondary distribution system

One-pipe, series loop system*

Primary-secondary system*

FIGURE 19-4 _____

Balancing valve used to control the flow of water in a branch or secondary loop of a distribution system

FIGURE 19-5 _____

Two-pipe, direct return supply and return system

Two-pipe, direct return system*

Two-pipe, reverse return system*

Three- or four-pipe system*

multiple loops can be used. In this case, each loop is treated as a separate piping system.

This system requires a method of ensuring a flow of water through each of the terminal devices and a way to balance that flow with the heat demand for the particular load. To accomplish this purpose, automatic or manual balancing valves are installed on either the supply or the return for each individual unit. Such a valve is shown in Figure 19-4.

Two-Pipe, Direct Return System

On larger jobs, especially when the designer desires to keep the pipe sizing to a minimum, a two-pipe system should be used. A **two-pipe, direct return system** incorporates a supply pipe and a return pipe. The return pipe runs parallel with the supply line but returns the water directly back to the boiler from each terminal device. Figure 19-5 illustrates a typical two-pipe, direct return supply and return system.

The water from the first unit out of the supply pipe is the first in to the return and first back to the boiler. In this system, the return water is running in the opposite direction to that in the supply line.

Two-Pipe, Reverse Return System

In a **two-pipe, reverse return system,** the main supply pipe runs from one heating unit to another. The return pipe runs parallel with it from the first unit on. See Figure 19-6 for an example of this system.

The hot water is taken from the supply line, through the secondary piping to the unit, and then to the return pipe from the outlet of the unit. The water flow in the return pipe then continues in the same direction as that in the supply pipe. This system makes it possible to decrease the size of the supply pipe as it goes from unit to unit. The return pipe size increases as it goes around the loop. The heating units are sized for only the quantity of water required for their individual load since they are actually in parallel with the main piping.

This system is the more desirable of the two-pipe systems because all of the water in the system flow entirely around the system and the friction in the piping is always the same for each loop.

Three- or Four-Pipe Systems

On a large job where the hot water is distributed for heating and chilled water is used for cooling in the building, a three- or four-pipe system can be used. In either a **three-** or **four-pipe system,** one supply pipe is used for carrying the heating water and a different pipe for cooling water. One or two return lines can be used. These systems normally are used only on jobs requiring heating in some zones at the same time as cooling is needed in others. See Figure 19-7 for an illustration of how a three-pipe system is used for heating and cooling at the same time.

Piping Arrangements

When a hydronic system is designed, the routing and location of supply and return water piping is an important consideration. In most buildings, it is important to locate the piping so that it is concealed; yet various components of the system will require service. Access is needed to these components. The piping used is generally fairly heavy and needs to be located so that it can be supported by the structure of the building. Designers must consider all of these points when locating the terminal units so the piping will fit into the general design of the building.

19.6 CIRCULATING PUMPS

In a typical hydronic system, pumps circulate the water through the various piping loops. If a system has more

FIGURE 19–6 _____
Two-pipe, reverse return system

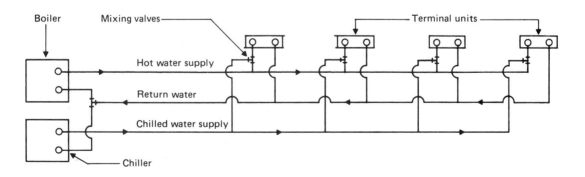

FIGURE 19–7 _____
Three-pipe supply and return system used on a heating-cooling application

FIGURE 19–8 _____
Typical line-mounted circulating pump used on a hydronic system

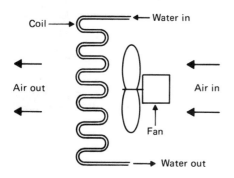

FIGURE 19–9 _____
Water-to-air heat exchanger

Centrifugal pump*

Feet of head – Term related to water pressure; pressure of a fluid expressed in terms of the height of a column of the fluid that it would support.

Water-to-air heat exchanger*

Radiator*

Convection coil*

than one zone loop, each loop will have a pump. Such a system is used when a building has separate zones, and each zone needs individual control.

The pumps used are centrifugal pumps. In **centrifugal pumps,** water is drawn in at the center of a housing in which an impeller wheel is rotated at high speed. The water is thrown by centrifugal force to the outer edge of the wheel. A discharge port is placed on the housing at one point on the outer perimeter. The water is forced through the port by the wheel.

Many types of circulating pumps are available. They are designed by physical arrangement as related to the way they are connected to the water pipes and by size related to the volume of water they will pump; some of the physical arrangements are line mounted or floor mounted. Figure 19-8 illustrates a typical line-mounted circulating pump.

Capacities of such pumps are given in gpm of water that can be circulated against a specified resistance to flow. The resistance is measured in **feet of head.**

19.7 TERMINAL UNITS

Heat exchangers must be employed to use the heat in the water in a hydronic system. The heat exchangers extract heat from the water and transfer the heat to the air. The air is then used for the heating medium. Figure 19-9 illustrates a **water-to-air heat exchanger.**

The heat exchanger in the hydronic system is generally in the form of a **radiator** or a **convection coil.** In both the radiator and the convection coil, the hot water flows through the inside of the elements and air is heated as it passes over the outside of the element.

In some cases, coils are built in to the walls, floors, or ceilings of a building. The building surfaces become a source of radiant heat. More often, the coil is installed in a cabinet located in the space to be heated. Whatever method is used for the final exchange of heat from the water to the air, the device used to make that exchange is called the terminal unit of the system. Figure 19-10 illustrates some typical terminal devices used for hydronic systems.

Selection Criteria

Some basic design considerations go into the selection of the type of terminal unit to use for a particular job. These considerations are

1. Aesthetic requirements
2. Physical limitations
3. Heat distribution requirements
4. Sensitivity of control requirements
5. Piping arrangements
6. Ventilation air requirements

There may be other considerations for special applications, but the factors in the preceding list should be considered for every job.

Aesthetic Requirements.   If the terminal device is being used in a warehouse or a manufacturing area or any part of a building where appearance is not important, relatively simple equipment can be used. Either direct radiation or unit heaters could be chosen. **Radiation panels** or **unit heaters** usually are suspended from the ceiling or some overhead part of the building structure. Because appearance is not important, the piping and control valves can be exposed. The designer's main concern is to get the units high enough so they will be out of the way, yet low enough and in the right locations so the warm air will be directed where it is needed.

In areas where appearance is important, such as offices, the terminal devices are usually chosen to blend in to the surroundings. Coils can be enclosed in cabinets that are neatly finished and painted a neutral color. These cabinets fit along the walls, either close to the floor or low enough on the wall to be inconspicuous. Figure 19-11 illustrates an enclosed convection unit of this type. The unit is both attractive and functional.

School rooms and large office spaces commonly have terminal units that fit under the windows. This arrangement makes it possible to bring in ventilation air through the outer wall. Heat is introduced to the room at its perimeter, where the heat loss is the greatest. This type of unit is called a unit ventilator. The units are normally

Radiation panel – Heating panel that emits radiant energy.

Unit heater – Self-contained heating unit, with a heat exchanger and blower in one cabinet, used with ductwork.

Baseboard convection unit

Unit ventilator

Convector

FIGURE 19-10 _____
Selection of terminal devices used with hydronic heating systems

FIGURE 19–11
Enclosed convection unit

enclosed in metal cabinets and painted to match the rooms in which they are used.

Physical Limitations. Most buildings are not designed with the placement of the mechanical equipment in mind. Designers must not only put the equipment in the best place for heating the building but be sure the equipment will fit. In most cases, the best location for the terminal units is along the outside walls, since that is where the greatest heat loss occurs. If the general arrangement of the rooms makes such placement impossible, then the designer should fit the units where they will do the best possible job of heating the space and still fit into the space available.

Heat Distribution Requirements. A designer must bear in mind that the space actually will be heated by air, although the heating medium is water. The terminal units are simply devices that use hot water to heat air, which is then circulated through the space to be heated. The terminal devices, or heating units, should be located so the heated air will circulate through the space as evenly as possible and, especially, should cover any areas of high heat loss, such as windows, doors, and/or outside walls, with a blanket of warm air.

Sensitivity of Control Requirements. A hydronic heating system provides one of the most important functions of a comfortable heating system—that is, a source of heat that remains constant. This source of heat is the mass of hot water in the system. After the water is heated, it gives up heat relatively slowly. This slowness can be a disadvantage when the temperature in the building changes rapidly, as it might in a building where the number of people varies or in a building where there are large windows on the south or west side. Because of the hot water in the system, heat can radiate from the terminal units after the thermostat in the room or zone where the units are located is satisfied. Because of the mass of water in the system, a hydronic system is relatively slow to provide heat. To compensate for this slowness of response, controls should be used on the components of the system that can sense changes in temperature rapidly and that can change the water flow in the units and the air flow through them as needed. These specific controls are covered in general later in this chapter and in detail in Chapter 20.

Piping Arrangement. The particular piping arrangement to be used in a building must be considered when selecting the terminal units. The location of the main pipes in the system may make it impossible to use certain types of units. The general layout of the piping system will certainly influence the location of individual terminal units. Details concerning the terminal units and their selection are found in Chapter 20.

Ventilation Air Requirements. One of the prime requirements for comfort in a building is that the air be kept fresh by the addition of clean outside air. Many of the types of terminal units used with hydronic systems are basically radiant heat units. These units have no provision for the introduction of ventilation air to the building through the unit. The designer should make the terminal unit selection with the above facts in mind. If the building will have a fairly large people load, then the terminal units should be a type that will introduce ventilation air into the rooms which the people normally occupy. Unit ventilators are one such type. If the building will be used primarily for storing goods, and people are not a big part of the load, then units that merely heat the air without introducing ventilation air are acceptable.

Types of Units

Several general types of terminal units, and many variations of each type, are available. The main types are

1. Direct-radiation elements
2. Radiators
3. Convection coil units

All of these types are used to some extent, and each has advantages and disadvantages. Each is discussed in the following sections.

Direct-Radiation System. A **direct-radiation system** is one in which the heating elements (in this case, hot water coils) are built into the walls, floors, or ceilings of a building. The coil warms that part of the construction, and the hot surface radiates heat into the space. The heat travels in the form of rays. When these rays strike any object in the room, the radiant energy becomes heat energy.

Direct-radiation system*

FIGURE 19–12
Cast-iron radiator

Bare copper coil*

Bare steel coil*

Finned-tube coil*

Duct heater*

Blower coil*

Convector*

Forced air convector*

This form of heating has some disadvantages. The surface temperature must be fairly high to emit enough rays for warmth; therefore, a lot of coil surface is required. A second disadvantage arises from the coil location; the coil is embedded in the building structure. If a leak occurs, the structure usually is damaged before the leak is found. Finally, the piping is difficult to get to for repairs.

Radiation systems are used in some locations, but such systems will not be covered further in this text.

Radiators.   When hot water heating systems were first used, they usually employed only radiators as a means of heating the air in a building. Cast-iron radiators are an example. See Figure 19-12.

The radiators heat the air by convection, direct radiation, and natural draft. The warm air moving across the surface of the radiator is the only air mover in the system. Simple radiator systems are seldom used in hot water systems today, except in commercial or industrial applications. Circulated air systems are more common.

Convection Coil Units.   The basic component used in most hydronic heating units to heat the air is the convection coil. When a convection coil is used, air is forced across the coil to carry the heat away from the coil and into the space to be heated, as shown in Figure 19-13.

Two basic types of coils are used. The first is a **bare copper** or **steel coil.** The size of the tubing varies, depending upon the amount of heat transfer required. The air is simply blown over the coil in a direction at right angles to the tubing. The amount of heat extracted from the water within the coil is dependent upon the water temperature, the thickness of the coil material, the metal in the coil, and the amount of air moving across the coil.

The second type of coil used, and the more common one for comfort heating systems, is the **finned-tube coil.** Figure 19-14 illustrates a finned-tube coil. Aluminum fins are pressed onto the tubing in such a way and increase the ability of the coils to heat the air passing over them. The coil is still installed in the airstream so the tubes are at right angles to the air flow, but the aluminum fins are parallel with the flow. See Figure 19-15.

Duct Heaters.   If a central air handling system is used in which air is distributed to various parts of a building

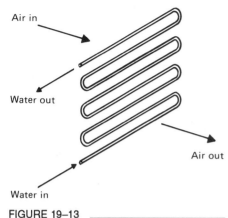

FIGURE 19–13 _____
Bare coil heating unit with air as the heating
medium

FIGURE 19–14 _____
Finned-tube coil

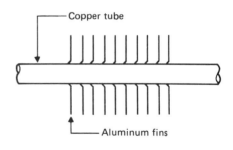

FIGURE 19–15 _____
Fins pressed on the coil to ensure good heat
transmittal

through ductwork, coils can be installed in the ducts so air
from the central blower passes through the coils. Such
coils are called **duct heaters.** A diagram of a duct heater
application is shown in Figure 19-16.

Blower Coils.   Sometimes it is necessary to furnish heating in areas that are not provided with air from a central
source blower. In this case, blower coils are used. A **blower coil** unit is a self-contained blower and heating coil
installed in a common cabinet. Figure 19-17 illustrates a
typical blower coil unit. Blower coil units usually have the
controls necessary for their operation as part of the package. Quite often, such units also have filters in the cabinet.

Convectors.   A **convector** is a unit with a heating coil
encased in a metal cabinet. A return air inlet is located at
the bottom of the unit, and a warm air outlet is located at
the top. See Figure 19-18 for a typical convector unit.

Cooler air enters at the bottom of the unit. The air is
heated by the coil and rises. The motive power, which
moves the air upward and out of the convector, is provided by the natural draft effect of the rising warm air. Some
convectors, called **forced air convectors,** contain blowers
to move the air.

The outer metal cabinet on the convector improves the
draft effect and the appearance of the unit. The cabinet
also helps direct the air up.

FIGURE 19–16
Diagram of a typical duct heater application

FIGURE 19–17
Diagram of a blower coil unit

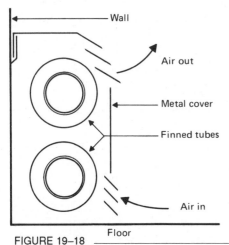

FIGURE 19–18
Typical convector unit

19.8 CONTROL SYSTEM

The last major components in a hydronic system are the control systems. A hydronic control system has two parts. The first part includes those control components that control the burner and maintain water temperature in the boiler; the second part controls the operation of the terminal devices. There are many different combinations for each part of the system and also many ways to interlock the two.

The general scheme and operation of a control system has already been covered in Chapter 10. Some general observations are given here concerning the special characteristics of hydronic control system design.

Because of the large amount of water used in a hydronic system, it reacts relatively slowly to temperature changes within a building. To compensate for this slowness, controls should be used that anticipate changing loads as much as possible. The usual method is to use outdoor reset controls that raise the water temperature in the boiler when the temperature of the outside air goes down.

Low voltage controls are recommended on the terminal devices. Such controls are more sensitive to temperature changes, and can indicate changes more quickly than an unanticipated system can. When rapid load changes are anticipated, as in a building with a lot of window area or shifting people loads, the controls should always be chosen to control airflow across the coils, as well as the flow of water in the coils.

19.9 SUMMARY

A hydronic heating system is one in which the heat for a building is generated in a central location and then is carried to different areas in a water system to heat separate parts of the building.

The major components of a hydronic heating system are the boiler, distribution system, circulating pumps, terminal devices, and the controls. Two basic types of boilers are used. These are described in Chapter 10. Several types of distribution systems are used. These types are usually described by the arrangement of the supply and return pipes and by the way the terminal devices are connected in the system. Pumps are used in pipe loop for circulation of the water through the system and are described by the method used for mounting them in the system and by the volume of water that they can distribute. The control system has two parts. One part controls the burner and regulates the temperature of the water available to the system. The second part controls the terminal devices and regulates the heat output of each.

19.10 TERMS TO KNOW

distribution system
one-pipe, series loop system
primary-secondary system
two-pipe, direct return system
two-pipe, reverse return system
three- or four-pipe system
centrifugal pump
feet of head
water-to-air heat exchanger
radiator
convection coil
radiation panel
unit heater
direct-radiation system
bare copper coil
bare steel coil
finned-tube coil
duct heater
blower coil
convector
forced air convector

CHAPTER

20

Hydronic System Design and Component Sizing

20.1 INTRODUCTION

For a hydronic heating system to operate, all parts must be sized correctly, and each part must work properly with the others. This chapter discusses the procedures to follow when designing a system: sizing the terminal devices and the supply and return pipes; calculating the total system friction; selecting the circulating pump, piping accessories, and controls; and determining the general layout of the equipment room.

20.2 SYSTEM DESIGN

The general layout of a total hydronic system and a description of the individual parts were described and explained in Chapter 19. Systems may vary, but the general type of component for each part will be the same. Each part of a system should be chosen to provide the best possible performance in that system for the application. Manufacturers' specification sheets or catalogs are the best source of information for each separate component.

20.3 DESIGN STEPS

Certain steps should be followed when laying out a plan for a heating system. These can be called the design steps. They are

1. Obtain or draw a building plan
2. Figure the heat load
3. Show the loation of the boiler and terminal devices on the building plan
4. Make the piping layout on the building plan
5. Size all components
6. Select the control system
7. Show the details of the boiler room piping on the building plan

If each step is completed in sequence, the design process will go smoothly, and the finished product will operate properly.

The first step is to obtain a plan of the building in which the system is to be installed. The plan may be a simple one-line drawing or a complete set of plans. It should show the walls, partitions, windows, doors, and any cabinets or detail that will affect the heating system. The plan should be drawn to scale. All major dimensions should be shown on it. See Figure 20-1 for a sample of the drawing. The designer will use this plan to show the heating system, equipment, and piping arrangement.

CONSTRUCTION DETAILS:

1. WINDOWS — SINGLE GLASS, NO STORM SASH
2. DOORS — W/ STORM DOOR
3. WALLS — FRAME WITH 2" INSULATION
4. CEILING/ROOF - 4" INSUL. IN ATTIC
5. FLOOR — OVER CRAWLSPACE (UNHEATED ROOM)
6. INFILTRATION — WINDOWS WILL BE HORIZONTAL SLIDERS W/ WEATHERSTRIPPING
7. CEILING HEIGHT 8' 0"

FIGURE 20-1

Floor plan of a typical residence with all data recorded for the layout of a hydronic system

The next step is to figure the heat load for the building. Before the components can be selected for the system, the designer must know how much heat needs to be furnished to offset the heat loss of the building. Chapter 14 explains how to calculate heat load. Each individual part of a hydronic system is sized to provide the amount of heat required to each part of a building. The calculations must be accurate to ensure that the final design will be adequate. For design purposes, the heat load should be calculated for individual rooms or zones.

After the heat load is figured, the designer should write on the plan the load for each room. The terminal devices also can be selected and shown on the plan at this time, as shown in Figure 20-2.

Next, the boiler should be selected and its location shown on the drawing. If the boiler is placed on a floor other than the one that the terminal devices are on, then the location of the terminal devices should be shown on the same plan as the boiler. Figure 20-3 shows an overlay,

FIGURE 20–2

Floor plan from Figure 20-1 with the terminal devices shown for a simple hydronic layout

FIGURE 20-3
Boiler and terminal devices show on an overlay of the plan for Figure 20-1

made from the plan in Figure 20-1, with the boiler and the terminal devices located. In selecting the terminal devices and the boiler, as well as other mechanical parts of the system, the designer should refer to the manufacturer's specification sheets for the parts, which will give the necessary data for selection and sizing.

The location of the supply and return piping is next shown on the plan, in accordance with the system selected as best for the job. The pipe also can be sized at this time, using the gpm required for the terminal devices to determine the flow rate for each section of pipe. This step will be described in detail later in this chapter. Figure 20-4 shows the overlay with the piping layout on it.

After the piping layout is made and the piping is sized, all of the other components of the system can be selected and a detailed drawing made of the boiler piping and accessories. Figure 20-5 is a schematic drawing showing some of the details that should appear on this drawing. Details concerning the boiler are found in Chapter 10 in this text. The distribution systems were described in Chapter 29. Sizing of the piping and the other components of the system will be covered in the rest of this chapter.

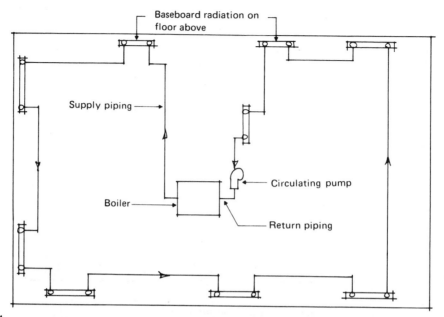

FIGURE 20–4
Supply and return pipe locations shown on overlay of the plan for Figure 20–1

FIGURE 20–5
Schematic drawing of a boiler and some of the parts and components related to it

20.4 TERMINAL DEVICES

Selection of the type of terminal devices to use usually is made when the distribution system is decided upon. The two parts of the system are so dependent upon each other that they must be considered at the same time in the design process. The designer should select the terminal devices from those described in Chapter 19 and show the location of each one on the drawing.

After the type of terminal device has been selected, it is necessary to select the proper size of each device. The heat load of each space should already have been figured, on a room-by-room basis, and the loads shown on the floor plan of the building. The location of the terminal devices should also be shown on the plan. This plan will be used during the rest of the design procedure and also later for balancing the system.

gpm per Terminal Unit

The next step in sizing the terminal devices is to convert the Btu/h required per unit to gpm per unit. The equipment, piping, valves, and other components used in the system are all sized for handling water. The water in the system will be flowing at a rate measured in gpm, so the loads to be handled by each terminal device should be converted to gpm. The following formula can be used:

$$gpm = \frac{Btu/h}{60 \times 1 \times 8.345 \times TD}$$

where

gpm	= amount of water to be circulated
Btu/h	= amount of heat to be delivered
60	= min. in an hr
1	= specific heat of water
8.345	= lb of water per gal
TD	= temperature drop of the water

In use, the divisor $60 \times 1 \times 8.345$ is constant. Its value of 500.7 is close enough to 500.0 so it can be rounded off and 500 used as a constant value. The formula then becomes:

$$gpm = \frac{Btu/h}{500 \times TD}$$

The use of this formula makes figuring the gpm required for any Btu/h fairly easy and is accurate enough for practical design use.

To make the conversion, it is necessary only to select the proper TD expected through a terminal unit and use the formula to determine the gpm for that unit. For design purposes, a TD of 20°F through any major part of the system can be used. This TD will have to be verified when the equipment is selected, but it is close enough for preliminary design purposes. Also for design purposes, it can be assumed that the initial supply water temperature will be 200°F and the leaving water temperature would be 180°F.

EXAMPLE 20-1

A zone in a building has a heat loss of 5568 Btu/h, and the TD through a terminal unit to serve that zone is 20°F. Find the gpm to be circulated.

Solution

$$gpm = \frac{Btu/h}{500 \times 20°F}$$

$$= \frac{5568}{500 \times 20}$$

$$= 0.56$$

The quantity of water to be circulated is 0.56 gpm.

Some thought will have to be given to the type of piping system that will be used. If a series loop system is used, then the TD through each unit affects the entire system. If a primary-secondary system is used, the effect of each unit is much less.

Selecting the Terminal Units

The selection of the terminal units must be made by type to fit the application and also by capacity for the heating load. Sizing selection charts for each of the different types of units are furnished by the manufacturers of the units. Designers should refer to these charts when making a selection. The selection chart shows the Btu/h output of each unit at a given flow rate (gpm) and for a variety of TD. If a unit is to be sized to heat a room requiring a given Btu/h, then reference should be made to the chart for a unit with that Btu/h capacity, and the gpm and TD for the unit will be shown. Conversely, if a flow rate has been established for a particular part of a system and the Btu/h required is known, then the designer can use the gpm and Btu/h to see if the TD is acceptable.

20.5 PIPE SIZING

With the location of the supply and return piping shown on the plan, the next step is size the pipe for the amount of water that will have to flow through it. Since the water is the medium that carries the heat to the different parts of the system, proper sizing will ensure that the correct amount of heat is available where it is needed.

In sizing the pipe, the designer must consider the **friction loss** through the pipe, as well as the gpm required. The friction loss in piping is like static pressure drop in ductwork. Friction loss is an indication of the amount of resistance to the flow of the water in the pipe.

Water pressure can be measured in feet of head. This term indicates the amount of pressure exerted by a column of water of a given height, or conversely, the amount of pressure that would support a column of water of a given height. For instance, water pressure of 6 ft of heat would raise a column of water 6 ft in vertical height. **Static head** is another term for this pressure. See Figure 20-6.

Friction in a water pipe is measured in feet of friction head. This term relates to the amount of resistance in a given length of pipe. **Friction head** is measured in feet of head and is equivalent to the pressure reduction caused by the friction in the pipe. Figure 20-7 illustrates friction head. Friction head is measured in **mils per foot (mils/ft)** or **feet per hundred feet (ft/100 ft)** of pipe. The term "mils/ft" is actually milinches (1/100th of an in.) of head per ft of pipe; the term "ft/100 ft" is 1 ft of head per 100 ft of pipe.

Just as a duct system is designed for a given amount of friction per 100 ft of duct, so is pipe. An arbitrary selection of 300 mils/ft, or 2.5 ft/100 ft, will give a design that is right for most small- to medium-sized jobs. The resulting total friction loss usually will be within the range of a standard circulation pump.

A pipe sizing chart is used for the actual sizing of the pipe. The chart is similar to the equal friction chart used for ducts but has data for pipes. Different charts are used for different types of pipes because of the difference in the roughness of the inside surface. Table 20-1 shows a pipe sizing chart with pipe size as a function of the flow rate of water in gpm and the friction loss in mils/ft. To use the chart, the gpm required is plotted against the desired friction, and the pipe size is shown for the two. When they do not coincide on a pipe size exactly, then the next larger size pipe should be used. For example, if a section of pipe must carry 17.0 gpm of water at 300 milinches per foot, Table 20-1 shows that a 1.5 in. pipe should be used. A 1.5 in. pipe will actually carry 18.0 gpm at 300 milinches, but since it is the size closest to 17.0 gpm it should be used.

FIGURE 20–6 _____
Static head. The height in feet that the water has to be raised in a system is static head.

FIGURE 20–7 _____
Friction head and pressure difference. Friction head is the friction in a piping system given in equivalent feet of height and pressure difference is an indication of friction in the piping system.

Friction loss – Pressure drop caused by resistance to water flow in a pipe.

Static head*

Friction head*

Mils per foot (mils/ft)*

Feet per hundred feet (ft/100 ft)*

* Term defined in text

TABLE 20–1
Pipe Sizing Chart

	1/4	3/8	1/2	3/4	1	1¼	1½	2	2½	3	3½	4	5	6	8	10	12
						Nominal Pipe Size (in.)											
	Carrying Capacity of Various Pipe Sizes in Gallons per Minute*																
700	0.53	1.2	2.2	4.8	9.3	19.0	29.0	56.0	90.5	170.0	245.0	350.0	650.0	1,100.0	2,300.0	—	—
650	0.50	1.13	2.1	4.6	9.0	18.0	28.0	54.0	88.0	160.0	235.0	330.0	620.0	1,040.0	2,200.0	—	—
600	0.48	1.1	2.0	4.4	8.5	17.3	26.8	51.0	85.0	150.0	225.0	320.0	590.0	1,000.0	2,100.0	4,000.0	—
550	0.46	1.03	1.95	4.2	8.1	16.5	25.3	49.0	80.0	145.0	215.0	300.0	560.0	940.0	2,000.0	3,800.0	—
500	0.44	1.0	1.8	3.9	7.7	16.0	24.0	47.0	76.0	140.0	210.0	290.0	530.0	900.0	1,900.0	3,600.0	—
450	0.41	0.94	1.7	3.7	7.3	15.0	22.7	44.0	73.0	131.0	190.0	270.0	505.0	840.0	1,800.0	3,400.0	—
400	0.39	0.87	1.6	3.5	6.9	14.0	21.2	41.0	68.0	123.0	180.0	250.0	480.0	800.0	1,700.0	3,200.0	—
375	0.37	0.85	1.55	3.4	6.6	13.5	20.6	40.0	65.0	120.0	175.0	245.0	460.0	760.0	1,600.0	3,100.0	—
350	0.36	0.81	1.5	3.2	6.3	13.0	20.0	38.5	63.0	115.0	165.0	235.0	440.0	730.0	1,550.0	2,950.0	—
325	0.35	0.80	1.45	3.1	6.1	12.5	19.0	37.0	60.0	110.0	160.0	225.0	420.0	700.0	1,500.0	2,850.0	—
300	0.33	0.75	1.38	3.0	5.8	11.8	18.0	35.0	57.0	104.0	150.0	215.0	400.0	670.0	1,400.0	2,700.0	—
275	0.31	0.72	1.3	2.9	5.6	11.3	17.2	33.5	55.0	100.0	145.0	205.0	380.0	640.0	1,350.0	2,600.0	—
250	0.30	0.68	1.25	2.7	5.3	10.9	16.5	32.0	52.3	95.0	138.0	195.0	360.0	600.0	1,300.0	2,450.0	—
225	0.28	0.65	1.18	2.6	5.0	10.2	15.5	30.0	49.0	90.0	130.0	182.0	340.0	570.0	1,200.0	2,300.0	—
200	0.27	0.60	1.1	2.4	4.7	9.5	14.5	28.0	46.0	83.0	122.0	170.0	320.0	530.0	1,200.0	2,300.0	4,000.0
175	0.25	0.57	1.01	2.25	4.4	8.9	13.5	26.0	43.0	78.0	113.0	160.0	300.0	500.0	1,100.0	2,150.0	3,800.0
150	0.23	0.52	0.96	2.07	4.0	8.2	12.5	24.0	38.5	71.0	104.0	145.0	275.0	450.0	1,050.0	2,000.0	3,500.0
125	0.21	0.48	0.85	1.85	3.6	7.4	11.2	22.0	36.0	65.0	95.0	132.0	247.0	410.0	950.0	1,800.0	3,250.0
100	0.18	0.42	0.77	1.64	3.2	6.5	10.0	19.0	31.0	56.0	83.0	117.0	217.0	360.0	850.0	1,650.0	3,000.0
75	0.16	0.36	0.64	1.4	2.7	5.5	8.5	16.4	28.0	48.0	72.0	100.0	182.0	310.0	750.0	1,450.0	2,700.0
50	0.13	0.28	0.51	1.13	2.2	4.5	6.7	13.0	21.5	39.0	56.0	80.0	152.0	245.0	500.0	980.0	2,000.0
																	1,550.0

Note: The pipe is sized to carry water at a given flow rate and at a fixed friction loss.
Source: Bell & Gossett

Sizing Procedure

The gpm required to each terminal device should be shown on a heating unit selection chart and also on the layout of the building. A typical heating unit section chart is illustrated in Figure 20-8.

If the chart was used during the selection process for the terminal units, the gpm will already be listed on it. If it was not, the data should be entered now. The other columns can be used for data relating to the pipe sizing.

The next step is to size the pipe. The following steps should be taken:

1. Figure the gpm needed in each section of pipe
2. Select the friction heat to use
3. Find the pipe size on the pipe sizing chart
4. Record the pipe size on the unit selection chart and on the plan
5. Indicate on the plan where the pipe size transitions take place

Heating Unit Selection				
Space	Btu/h	gpm	TD	Unit selected

FIGURE 20–8

Heating unit selection chart used to record the data for proper selection of the terminal devices for a given job

Sizing a One-Pipe, Series Loop System

A one-pipe, series loop system is simple to size since the flow rate (gpm) is the same all around the loop. Using a friction head of 300 mils/ft of pipe, the designer uses the pipe sizing chart to find the pipe size for the desired gpm. The total gpm for the loop should have been determined when the terminal units were selected. If it was not, it can be found at this point.

EXAMPLE 20-2

If a one-pipe, series loop is to be sized for a job that requires a total of 28,250 Btu/h of heating, with a 20°F TD through the system, regardless of the number of terminal units, the gpm for the Btu/h must first be found. This is done by using the following formula:

$$\text{gpm} = \frac{\text{Btu/h}}{500 \times \text{TD}}$$

Then the pipe is sized for the gpm from the pipe sizing table.

Solution

$$gpm = \frac{Btu/h}{500 \times TD}$$

$$= \frac{28,250}{500 \times 20}$$

$$= 2.825$$

The quantity of water is 2.825 gpm. The pipe is then sized by referring to the sizing chart, which shows that a 3/4 in. pipe will carry 3 gpm at 300 mils/ft. Since this is the next size pipe above that for 2.825 gpm, then that is the size pipe to be used.

Because all of the water in a one-pipe, series system flows through ⅃ sections of the pipe, the loop pipe will be the same si: all around the loop. This size should be marked on the pipe layout on the floor plan.

Sizing a Two-Pipe System

A better example of pipe sizing can be shown by using a two-pipe, reverse return system. In this system, the terminal units are in parallel with the supply piping. The return pipe is separate from the supply but also in parallel with it. See Figure 20-9 for a diagram of a two-pipe, reverse return system. The water quantity in the supply pipe diminishes as it flows through the system, but the flow rate increases in the return line as it goes around the loop.

To size the pipe, the initial flow rate is used for the first section of pipe. This rate should be the total required for all units in the system. The pipe sizing is taken from the sizing chart, as explained for a one-pipe system. After the first unit, the flow rate through the supply pipe will be less because of the water taken out at the first unit, so the flow rate between the first and second unit is now used for sizing the pipe in that section. The same procedure is followed around the loop, sizing each section of pipe for the flow rate in that section, until all of the supply pipe has been sized.

The return pipe is sized in the same way, except that the pipe will increase in size as it goes around the loop. A

FIGURE 20–9 _____

Diagram of a two-pipe reverse return system showing the Btu/h load and the gpm requirements for each terminal unit

Pipe Sizing Data Chart			
Pipe	Designation	gpm	Size at 300 mil/ft
SUPPLY	A-B	5.74	1" ⌀.
DO	B-C	4.54	1" ⌀.
DO	C-D	3.58	1" ⌀.
DO	D-E	2.53	3/4" ⌀.
DO	E-F	1.40	1/2" ⌀.
RETURN	G-H	1.20	1/2" ⌀.
DO	H-I	2.16	3/4" ⌀.
DO	I-J	3.21	1" ⌀.
DO	J-K	4.34	1" ⌀.
DO	K-L	5.74	1" ⌀.

FIGURE 20-10
Pipe sizing data chart with data recorded for Example 20-3

pipe sizing data chart, in which the sections of pipe are identified by letters or numbers and the gpm is listed for each section, is invaluable for sizing pipe in any system where different flow rates occur in the different sections of the system. Figure 20-10 shows such a chart with data on it, for Example 20-3, which follows next.

EXAMPLE 20-3

Find the pipe sizing for the job shown in Figure 20-9.

Solution

First, find the gpm through each unit so the gpm for each section of pipe can be figured. To do this calculation, the total system TD should be decided upon and then the total system gpm found. If a total TD of 20°F is chosen, then the total flow rate is found by the following formula:

$$gpm = \frac{Btu/h}{500 \times TD}$$

$$= \frac{57{,}400}{500 \times 20}$$

$$= 5.74$$

The flow rate through each terminal unit can be found by proportioning the 5.74 gpm for the total heat load to each unit by unit heat load. The results are shown on the heating unit selection chart, Figure 20-11. The total TD can also be proportioned per each unit. The result should be shown on the chart.

The sections of pipe are identified by letter or number as on the plan. See Figure 20-12. The sections are also as listed on the pipe sizing data chart, Figure 20-10.

The gpm through each section of pipe is figured by starting with the full flow through the first section of pipe (A-B) and then deducting the gpm as the water is taken out at each terminal, as recorded on the pipe sizing data chart. The gpm of the return pipe is figured in the same way, but the water is added as it comes out of each terminal unit. This information also is shown on the pipe sizing data chart.

Heating Unit Selection				
Space	Btu/h	gpm	TD	Unit selected
A	12,000	1.20	4.18	
B	9,600	.96	3.34	
C	10,500	1.05	3.66	
D	11,300	1.13	3.94	
E	14,000	1.40	4.88	
TOTALS	57,400	5.74	20°F	

FIGURE 20-11 _____

Heating unit selection chart with data recorded for Example 20-3

FIGURE 20–12

Sections of pipe identified with letters to match those on the pipe sizing chart shown in Figure 20-10

FIGURE 20–13

Special valves used to balance the water flow in the different sections of the pipe

Finally, the pipe sizes are selected from the pipe sizing chart. The sizes are recorded on the pipe sizing data chart and also on the plan.

Next, the sections of pipe feeding each terminal unit are sized. Since the largest gpm is 1.40, and a 1/2 in. pipe will provide enough flow for that gpm, the 1/2 in. pipe can be used for all of the supply and return branches to the terminal units.

The final step is to show by use of a symbol (X) on the pipes where the transition in pipe sizes takes place. This marking is shown on the plan.

Sizing Fittings

The selection of balancing valves, shutoff valves, and other fittings for the distribution system usually can be made on the basis of the pipe sizing to the units. In some cases where special flow valves are used for automatic balancing, however, the flow rate through the valves also is required to make the selection. See Figure 20-13 for an illustration of some of these special valves. The flow rate in this case would be that required for the unit being provided water by the valve. The selection data are found in the valve manufacturer's specification sheets.

Other fittings, such as elbows and tees, are normally sized the same as the pipe to which they are connected. The additional pressure drop caused by these fittings is added to the total pressure drop of the system when sizing the pumps.

20.6 TOTAL SYSTEM FRICTION

The total system friction must be determined for the selection of a circulating pump. The pump will have to move the amount of water required for heating the building against the friction in the piping.

Total system friction is simple to calculate after the pipe has been sized for a given friction head per given length of pipe. In Examples 20-2 and 20-3, the friction head was to be 300 mils/ft, which is the equivalent of 2.5 ft/100 ft. To find the total friction in the piping in the system, it is necessary only to find the length of the pipe in the system, multiply the length by 2.5 ft/100 ft, and divide by 100 ft. The following formula is used:

$$\text{friction loss} = \frac{\text{length of pipe} \times 2.5}{100}$$

To find the total friction head in the system, the friction loss through any other components in the system, such as terminal devices, valves, or fittings, must be added to that for the piping.

Single-Pipe Systems

To size the piping for a single-pipe, series loop system, the length of the pipe loop should be scaled off the plan, or otherwise determined, divided by 100, and the resulting number multiplied by 2.5 ft/100. This calculation gives the friction loss in the piping only. To find the total friction loss for the system, the pressure drop through each terminal device should be added to the piping loss. This figure is used to size the circulating pumps.

Two-Pipe Systems

The total friction loss through a two-pipe system is figured similarly to that for a one-pipe system, with one difference: the friction loss through each unit is not added. The water flows through one device only in one circuit around the system. The friction loss through the device having the greatest loss is added to the friction loss for half of the supply pipe length and half of the return pipe length. This calculation will give the total friction drop for the system, disregarding the loss for fittings. For small jobs and preliminary figures, the total is close enough to use for sizing pumps.

This method gives the proper steps to follow for sizing any piping system for a typical loop system. The method should be followed for each loop, whether the system has only one simple loop or multiple loops for different zones.

20.7 CIRCULATING PUMP

After the total system friction is determined, the next step is to select and size the circulating pump, or pumps, for the system. Pumps are selected and sized to circulate the amount of water required in the system, against the friction, or resistance to flow, in the pipes. The flow rate is that determined for the total system when the pipe was sized. The resistance is measured in feet of water. The pump is selected to provide the desired gpm at that friction.

The total amount of water to be circulated already should have been calculated and entered on the pipe sizing data chart and the plan. This amount was calculated on the basis of the Btu/h required in each space when the terminal units were selected. The TD of the water and the design characteristics of the terminal units determine the total amount for each unit. Figure 20-14 is a photograph of an installation with two circulating pumps and some piping, valves, and accessories.

The total friction loss through the system also has been calculated. It is the friction loss through the pipe itself, plus the loss through the terminal units and the fittings in the system.

A circulating pump is selected by using a pump selection chart. This chart is furnished by the pump manufacturer. It shows the gpm delivery rate of the pump, as compared to the head in feet of water against which the pump must work. See Figure 20-15 for a pump selection chart.

The term **head in feet of water** is indicative of the resistance to water flow in the pipe, as measured by the

Heat in feet of water*

FIGURE 20–14

Installation with two circulating pumps, piping, valves, and accessories feet of water in a vertical column. A designer should always use a chart that has been developed for the particular pump under consideration.

EXAMPLE 20-4

Using the pump selection chart shown in Figure 20-15, select a pump for a job requiring a 20 gpm flow rate in a particular loop against a head of 8 ft.

Solution

Locate 8 ft of head on the vertical scale on the left of the chart. Then move across to the right on the horizontal line. Where the horizontal line intersects a vertical line coming up from the 20 gpm point on the flow rate scale at the bottom of the chart, look for the curing line related to the capacity of the pump. In this example, using the chart in Figure 20-15, the pump called for is a series PR booster

How to select a B&G Booster Pump
Required: 10 GPM at 6 ft. head. Look first at the bottom of the Booster Pump Capacity Chart where pump delivery is shown. Run a line straight upward from the 10 gallon point until it intersects a horizontal line from the 6 ft. head on scale at left. The nearest pump curve, or one slightly above this intersection is the proper selection. The Series 100 Booster pump should be used. It is not advisable to select a Booster pump with a head under 2½ ft.

FIGURE 20–15
Circulating pump selection chart from a manufacturer's catalog

pump. This pump would be the proper selection for this particular job.

If this were a multiloop job, the same procedure would be followed for each loop in the system. Individual pumps can then be individually controlled to give the proper temperature control to each zone in the building.

20.8 ACCESSORIES

In addition to the major components of the system, certain other components are necessary to a system for safe and proper operation. These components are called accessories to the other parts. The most important ones are: the **expansion tank**, the **relief valve**, **air vents**, and certain other valves, such as pressure-reducing and flow-control valves.

The expansion tank is used to accommodate expansion of the water in the system as the water is heated. Hot water has a greater volume than cold water. Because water in the liquid state is noncompressible, a great deal of pressure would be exerted if there was no allowance for it to expand as it was heated. The expansion, or compression, tank is filled with air and connected into the piping system in such a way that water will not flow into the tank naturally. Instead, the water will be forced into the tank as the water in the piping system expands. The air in the tank acts as a cushion. Since the air will compress, the water in the system can expand, but only at a controlled rate.

Expansion tank – Air tank located above the boiler in a hydronic system and plumbed into the system in such a way that the tank absorbs the extra volume of water when the water is heated.

A relief valve is always used on a boiler. The valve is spring loaded or weighted; it will open if the pressure in the system exceeds what is considered to be safe. If any malfunction of the controls causes the burner to stay on too long, the relief valve will open and prevent dangerous pressure build-up in the system. Most relief valves open on pressure build-up. Some also will sense high water temperature and open if either pressure or temperature exceeds a set limit.

Relief valve – Valve plumbed into the piping system above the boiler in a hydronic system that is designed to open and relieve the pressure in the boiler if that pressure exceeds a particular set-point.

As water is heated, air is released from it. This air naturally collects at high points in the piping system and prevents the water from flowing. Air vents are used to allow the air to escape into the atmosphere. These vents are installed in the piping system at the high points. Since the system cannot be left open or water would flow out, the valves have to be a type that will open only when there is

Air vents – Fittings installed at high points in the piping system of a hydronic system to vent air from the water circulating through the pipes.

air present but not water. One of two types is used. One type is manual. This valve can be opened by an operator when system operation indicates that air is present. The other valve is automatic. It will open by itself when air is in the line but will close as soon as it senses water.

Various other valves are used to control pressure or water flow. Among these are pressure-reducing valves to control the upper limit of the water pressure in the system, check valves used to make sure the water flows in one direction only, and gate and globe valves for flow control or shutoff control.

Other special flow-control valves can also be used in a typical system. The designer should refer to a plumbing diagram of the specific type of system to determine what valves and controls are needed. Manufacturers' specification sheets are the best source of detailed information concerning valves or controls.

20.9 CONTROLS

A hydronic control system is made up of two parts. One part controls the air temperature at the terminal device. This arrangement is necessary in a hydronic system because water is heated at one place in a building and then used to heat air in other parts of the building.

The main control that regulates the temperature of water in a boiler is called an aquastat. This device is a thermostat that senses water temperature and opens or closes an electric switch depending upon the temperature. An aquastat has a temperature scale and an adjustable indicator that can be set for a desired temperature, the setpoint of the system. If the temperature of the water goes below the setpoint, the electric switch closes. If the temperature goes above the setpoint, the switch opens. The switch is wired into an electric circuit that controls the burner. If the switch closes, the burner is actuated; if the switch opens, the burner is turned off. The burner cycles off and on to maintain the water temperature at or near the setpoint temperature.

To make the system better able to compensate for fluctuating loads, an outdoor reset is sometimes used in conjunction with the aquastat. This control will reset the setpoint temperature to compensate for changing temperatures outside the building. If the outdoor temperature goes down, the water temperature is reset higher. If the outdoor temperature goes up, the water temperature is set lower.

The second part of the control system (the part that controls the terminal devices) can be as simple as a balancing valve that is adjusted for a constant volume of water to flow. This type of control does not adjust to varying loads and so is practical only for the simple systems. If some control of the heat output of the device in relation to the load is needed, then a thermostatically controlled valve can be used to vary the flow rate according to the load. For better control of the heat output, air temperature controls are used with the water flow controls. This type of control will turn the blower off and on as the heating load varies.

In simple systems, the controls are mounted on the terminal devices themselves. For better control of the room air temperature, the thermostat is wall mounted. The thermostat is then connected to the motorized valves or blower control in the unit by an electric circuit.

20.10 EQUIPMENT ROOM LAYOUT

The final step in designing a hydronic heating system is to arrange the piping scheme that is to be used in the equipment room for connecting the boiler, expansion tank, control valves, and circulating pumps. This equipment usually is located in an equipment room in the building to be heated or adjacent to it in a special room for that purpose. The room itself should have been planned in the initial building plan. A typical piping layout for a small system equipment room is shown in Figure 20-16.

Consideration must be given to the location of many components necessary to complete the system. The designer should decide how best to fit these components into the space that has been provided in the building. Normally, the designer will find that the best method is to work on the floor plan of the equipment room first and then to use elevation and section drawings to show details.

Many different layouts are possible. The designer should select the one to be used for each job with care. The physical size of the equipment and components has to be considered, as well as the physical locations. Unfortunately, room is not often available to spread the equipment out too much.

FIGURE 20–16

Typical piping layout for small system equipment room

20.11 SUMMARY

Designing the hydronic system and sizing the components is similar to designing a warm air system. The design steps are

1. Obtain or draw the building plan
2. Figure the heat load
3. Locate the boiler and terminal devices on the building plan
4. Locate the piping layout on the building plan
5. Size all components
6. Select the control system
7. Show the details of the boiler room piping on the building plan

At this stage in the design process, selection of the proper type of components for the hydronic system should have been completed, as covered in Chapter 19. Now the components must be properly sized physically in order to fit into the building and also according to capacity in order to meet the heating requirements of the building.

In order to size the terminal devices, the heat load for each device must be converted from Btu/h to gpm. When the gpm requirement for each terminal device has been determined, manufacturers' sizing selection charts are used to select the correct devices for the job.

Next, the return and supply pipes are sized so that they will carry the proper amount of water for each part of the system they service, taking into consideration the resistance to flow of water in each section of pipe. The total system friction, or pressure drop, is then calculated to determine the amount of resistance that the distributing pump has to overcome. The distributing pump is sized taking into consideration the total amount of water to be circulated and the total system friction. Then, the accessory items, such as expansion tank, relief valve, and shutoff and balancing valves, are selected and sized. Finally, the arrangement of all of these parts in the equipment room is decided upon.

Boilers, circulating pumps, accessories, and controls are all selected and sized by using manufacturers' data sheets that give the specific details for each one. The components are selected to match each other and to provide the proper flow rate and control to heat the building according to the heat load.

20.12 TERMS TO KNOW

friction loss
static head
friction head
mils per foot (mils/ft)
feet per hundred feet (ft/100 ft)
heat in feet of water
expansion tank
relief valve
air vents

21.1 INTRODUCTION

Troubleshooting is the act of locating and identifying the source of trouble when a heating unit is not working properly. The investigation is preceded by the observation that something is wrong with the system, and it is followed by **service** to correct the problem.

The first requirement for troubleshooting is knowledge of how the equipment operates normally. The normal function of gas-fired, oil-fired, and electric furnaces is covered in the preceding chapters. In this chapter, the troubleshooting procedure is described. The chapter covers analyzing the complaints, identifying system components related to the complaint, determining what part is faulty, and checking the part to verify the problem.

21.2 ANALYZING COMPLAINTS

When a heating system in a building is not working properly, the building occupants usually complain. Normally the first indication that something is wrong is that the building air temperature may be too hot or too cold, that is, it varies from the setpoint temperature on the thermostat. Since the building occupant generally is not familiar enough with the heating system to identify the problem, he or she will make a complaint.

The complaint will be relayed as a general statement to the person who will do the troubleshooting, and this general statement will need some clarification as to specifics. The following is a list of some common complaints:

1. No heat
2. Insufficient heat
3. Too much heat
4. Fluctuating heat
5. Noise
6. Odor
7. Excessive cost of operation

A general description of each of these complaints follows. The troubleshooting steps to follow for each of them is given later in the chapter.

Troubleshooting*

Service – Corrective action taken to repair or adjust a malfunctioning unit so that it will work correctly.

* Term defined in text.

No Heat

This complaint is self-explanatory. When a building does not warm up at all when the thermostat is calling for heat the heating unit is probably not working. In this case, a troubleshooter must determine what is wrong.

Insufficient Heat

In some cases of heating unit malfunction, the heat output of the unit is reduced. The complaint will be that there is not enough heat. This complaint may not arise until the outside air temperature gets close to the design temperature. The condition may also occur intermittently at other outside temperatures. This condition indicates that the heating unit is either not performing correctly, or is undersized for the heating load.

Too Much Heat

This complaint is a cause for immediate concern. Too much heat may be caused by a unit that is not shutting off when the thermostat is satisfied, and can indicate a serious safety hazard. If a gas valve sticks open on a gas furnace, or the relays fail to open on an oil or electric furnace, overheating will occur. Ordinarily, the high temperature limits would shut the unit off, but in some rare cases, a limit may fail, and a very dangerous fire hazard exist. Overheating can be a dangerous condition.

Fluctuating Heat

In some cases of **malfunction,** the temperature in a building may fluctuate above or below the setpoint temperature on the thermostat. This problem can be caused by a faulty thermostat, or other operating control. A thermostat located on a cold outside wall or where cold drafts can affect it is one cause of fluctuating heat output. Other parts of the system may be at fault, such as supply register locations. A troubleshooter should check the entire system to determine the actual cause.

Malfunction – Not functioning correctly.

Noise

This complaint usually is created by mechanical failure of some part of the system or unit, but it may be caused by faulty ignition or operation of the burner. All noise complaints should be thoroughly investigated. They may indicate a failure that could lead to a more serious condition.

Odor

In the case of a fuel-fired unit, this complaint can indicate a very dangerous situation, such as leaking fuel or flue gas. A troubleshooter may have difficulty identifying the source of an unpleasant odor. Because of the danger created by escaping gas or incomplete combustion, the furnace should not be used until a complete check of its operation has been made.

Excessive Cost of Operation

This complaint is difficult to deal with because the cause can be so elusive. Perhaps there are problems with the operation of the unit; more likely, there are problems within the total system. If a heating unit runs more than it should there usually will be a complaint of overheating as well as excessive cost. The cost of operating any system is important. Because fuel costs generally are higher than most people would like, this complaint should be carefully considered.

21.3 TROUBLESHOOTING THE SYSTEM

A troubleshooter should always use a systematic approach to finding problems with equipment. Before troubleshooting starts, a complete understanding of the normal operation of a system is necessary. Next, a troubleshooter must obtain enough information about a system's operation from the person making a complaint, and from firsthand observation, to know how the unit is operating. Third, a troubleshooter must identify the general family of components in which the problem seems to lie. And fourth, the troubleshooter must identify the specific part or parts that seem to be faulty. Following these four steps is a systematic approach to troubleshooting.

To ensure that each step of a systematic troubleshooting process is followed, troubleshooting or service charts should be used. These charts show what steps to take as a troubleshooting analysis is made. The charts are a step-by-step procedure guide that lead from a complaint to the service procedures to be followed when a heating system malfunctions.

Understanding Heating System Operation

The first step in the troubleshooting process is understanding how a heating system is supposed to work. The basic operation of a heating unit is easy to understand, but the operation of the total system is more complex. In many cases, multiple components or parts of a heating system are tied together by the control system so they work together. Figure 21-1 shows a furnace and split system air conditioning unit connected by the control system to work as one unit. Two pieces of equipment may work together

HEATING AND COOLING APPLICATION WIRING DIAGRAM

FIGURE 21-1

Controls in the furnace and air conditioning units are often tied together by controls to function as one unit using common controls. An electronic cleaner often is used with a furnace. The cleaner control system affects the operation of the furnace through the furnace blower controls. If a malfunction occurs in the air cleaner control system, the furnace blower motor may also be affected. This could appear to be primarily a malfunction of the blower controls when in fact it isn't.

A general type of control system can also tie two or more components of a heating system together. Outdoor air temperature, damper system, or damper motor controls may all affect the operation of a heating system. Malfunctions in any part of the systems can, and often will, affect the operation of the heating system.

Recognizing Abnormal Operation

The second step in the troubleshooting process is observation of how the system is operating as it malfunctions. Before starting to look for problems, a troubleshooter should analyze any complaint received, and by observation of the unit, attempt to visualize what part of the system is malfunctioning. A few minutes spent in careful thought at the beginning of a troubleshooting call may save a lot of wasted effort. If the complaint is one of too much heat, for instance, then there is no use starting at the electrical disconnect and fuse box. Because the unit is working part of the time, there must be electric power to it. If, however, the complaint is that "nothing is working," then the place to start is at the disconnect.

Identifying Problem-Causing Components

The third step in the process is to identify the general components causing a problem. The major unit component categories are as follows:

1. Electric power components
2. Controls
3. Burner and heat-producing components
4. Air-handling components

These components and their parts are shown in Table 21-1 for the three main types of furnaces: gas-fired furnace, oil-fired furnace, and electric furnace. The letter and number designations of the parts relate to the troubleshooting chart (Table 21-2) and the procedure charts (Tables 21-3, 21-4, and 21-5). The letter designation G is for a gas-fired furnace, O is for an oil-fired furnace, and E is for an electric furnace.

TABLE 21–1 _____

Major Components and Parts of Gas-Fired Furnaces, Oil-Fired Furnaces, and Electric Furnaces

Gas-fired furnace	**Oil-fired furnace (continued)**
Electric power	Burner and heating components
1. Disconnect	O–9. Oil supply
2. Fuses (circuit breakers)	O–10. Piping
3. Control transformer	O–11. Pump
	O–12. Nozzle
Controls	O–13. Ignition transformer
G–4. Thermostat	O–14. Electrodes
G–5. Limit control	O–15. Combustion blower
G–6. Blower control	O–16. Fire pot
G–7. Blower relay	O–17. Heat exchanger
	O–18. Barometric damper
Burner and heating components	Air-handling components
G–9. Gas shut-off	19. Blower
G–10. Piping	20. Blower motor
G–11. Gas valve	21. Belt and drives
G–12. Thermocouple	22. Filter
G–13. Pilot safety	23. Outside air
G–14. Pilot burner	24. Ductwork
G–15. Burners	
G–16. Heat exchanger	**Electric furnace**
G–17. Diverter	Electric power
	1. Disconnect
Air-handling components	2. Fuses (circuit breakers)
19. Blower	3. Control transformer
20. Blower motor	
21. Belt and drives	Controls
22. Filter	E–4. Thermostat
23. Outside air	E–5. Limit control
24. Ductwork	E–6. Blower control
	E–7. Blower relay
Oil-fired furnace	E–8. Sequencers
Electric power	
1. Disconnect	Heating components
2. Fuses (circuit breakers)	E–9. Wiring
3. Control transformer	E–10. Elements
	Air-handling components
Controls	19. Blower
O–4. Thermostat	20. Blower motor
O–5. Limit control	21. Belt and drives
O–6. Blower control	22. Filter
O–7. Combustion safety	23. Outside air
controls	24. Ductwork
O–8. Primary control	

TABLE 21–2

Troubleshooting Chart for Gas-Fired Furnaces, Oil-Fired Furnaces, and Electric Furnaces

Components/Parts	Complaints						
	No heat	Insufficient heat	Too much heat	Fluctuating heat	Noise	Odor	High operation costs
Gas-fired furnace							
Electric power							
1. Disconnect	2						
2. Fuses (circuit breakers)	3						
3. Control transformer	9	3					
Controls							
G–4 Thermostat	1	1	1	1			
G–5 Limit control	10	2					
G–6 Blower control		4		2			
G–7 Blower relay					1		
Blower and heating components							
G–9 Gas shut-off	4						
G–10 Piping						1	
G–11 Gas valve	11		2				
G–12 Thermocouple	5						
G–13 Pilot safety	8						
G–14 Pilot burner	6						
G–15 Burners	7				2	2	
G–16 Heat exchanger		8				3	
G–17 Diverter						4	
Air-handling components							
19. Blower	12	5		3	3		
20. Blower motor	13	6			4		
21. Belt and drives					5		
22. Filter		7		4			1
23. Outside air				5		5	2
24. Ductwork							3
Oil-fired furnace							
Electric power							
1. Disconnect	2						
2. Fuses (circuit breakers)	3						
3. Control transformer	7	3			1		
Controls							
O–4 Thermostat	1	1	1	1			
O–5 Limit control	8	2					
O–6 Blower control		4		2			
O–7 Combustion safety controls	5						
O–8 Primary control	6		2				

TABLE 21–2 (continued)
Troubleshooting Chart for Gas-Fired Furnaces, Oil-Fired Furnaces, and Electric Furnaces

Components/Parts	Complaints						
	No heat	Insufficient heat	Too much heat	Fluctuating heat	Noise	Odor	High operation costs
Burner and heating components							
O–9 Oil supply	4	5				1	
O–10 Piping	9	6				2	
O–11 Pump	10				2	3	
O–12 Nozzle	11						
O–13 Ignition transformer	12						
O–14 Electrodes	13						
O–15 Combustion blower					3		
O–16 Fire pot						4	
O–17 Heat exchanger						5	
O–18 Barometric damper						6	
Air-handling components							
19. Blower		7		3	4		
20. Blower motor	14	8		4	5		
21. Belt and drives	15	9		5			
22. Filter				6			1
23. Outside air						7	2
34. Ductwork							3
Electric furnace							
Electric power							
1. Disconnect	2						
2. Fuses (circuit breakers)	3						
3. Control transformer	4						
Controls							
E–4 Thermostat	1	1	1	1			
E–5 Limit control	6	2					
E–6 Blower control		3		2			
E–7 Blower relay		4					
E–8 Sequencers	5	5	2		1		
Heating components							
E–9 Wiring	7						
E–10 Elements	8						
Air-handling components							
19. Blower	9	6		3	2		
20. Blower motor	10	7		4	3		
21. Belt and drives	11	8		5			
22. Filter				6			1
23. Outside air				7			2
24. Ductwork							3

TABLE 21-3

Troubleshooting Procedures for Gas-Fired Furnaces

Electric power components

1. Disconnect
   a. Check main disconnect for power
   b. Check unit disconnect for power (may be circuit breakers)
2. Fuses (circuit breakers)
   a. Check main fuses for opens
   b. Check unit fuses for opens (may be circuit breakers)
3. Control transformer
   a. Check for voltage on primary side
   b. Check secondary voltage
   c. Check for 60 cycle hum

Control components

G-4 Thermostat
   a. Check setting
   b. Check power input from transformer
   c. Do contacts close on call?
   d. Check location (sun, drafts, etc.)
   e. Is thermostat level on wall?
   f. Check anticipation

G-5 Limit control
   a. Is limit control closed in normal operation, and does it open at proper temperature?

G-6 Blower control
   a. Is blower control open when unit is cold, and does it close at proper temperature?

G-7 Blower relay
   a. Does blower relay close when fan switch is closed?

Burner and heating components

G-9 Gas shut-off
   a. Is shut-off valve open?

G-10 Piping
   a. Check piping for leaks

Burner and heating components (continued)

G-11 Gas valve
   a. Is gas valve closed when deenergized?
   b. Does valve open when energized?
   c. Check gas pressure at manifold

G-12 Thermocouple
   a. Does thermocouple produce adequate voltage?

G-13 Pilot safety
   a. Does pilot safety hold open when energized and close when deenergized?
   b. Does safety close both pilot and main valves?

G-14 Pilot burner
   a. Does pilot flame impinge properly on the thermocouple?
   b. Is burner clean?
   c. Check gas adjustment at pilot valve

G-15 Burners
   a. Is burner flame correct?
   b. Are ports clean?
   c. Are orifices correct size?
   d. Is there sufficient combustion air?

G-16 Heat exchanger
   a. Check for cracks or burn-out
   b. Is there any flame impingement?
   c. Check for sooting

G-17 Diverter
   a. Check for spillage

Air-handling components

19. Blower
   a. Is blower wheel clean?
   b. Check for correct rotation
   c. Check speed of rpm
   d. Is wheel noisy?

TABLE 21-3 (continued)
Troubleshooting Procedures for Gas-Fired Furnaces

Air-handling components (continued)
20. Blower motor
    a. Does motor operate when blower control closes?
    b. Is motor noisy?
    c. Does motor run hot?
21. Belt and drives
    a. Is belt broken or cracked?
    b. Check belt tension
    c. Are pulley sizes correct for cfm required?
22. Filter
    a. Is filter in place?

Air-handling components (continued)
    b. Does filter need cleaning or replacing?
    c. Is filter efficiency correct for job?
23. Outside air
    a. Check for proper quantity of outside air
24. Ductwork
    a. Is ductwork leaky?
    b. Is ductwork too small?
    c. Does duct run in uninsulated space?

TABLE 21-4
Troubleshooting Procedures for Oil-Fired Furnaces

Electric power components
1. Disconnect
    a. Check main disconnect for power
    b. Check unit disconnect for power (may be circuit breakers)
2. Fuses (circuit breakers)
    a. Check main fuses for opens
    b. Check unit fuses for opens (may be circuit breakers)
3. Control transformer
    a. Check for voltage on primary side
    b. Check secondary voltage
    c. Check for 60 cycle hum

Control components
O-4 Thermostat
    a. Check setting
    b. Is thermostat receiving voltage from transformer?
    c. Do contacts close on call?
    d. Check location (sun, drafts, etc.)
    e. Is thermostat mounted level on the wall?

Control Components (continued)
O-5 Limit control
    a. Are limit control contacts normally closed, and do they open at the proper temperature?
O-6 Blower control
    a. Are blower control contacts normally open, and do they close at proper temperature?
    b. Does sure-start fan switch close after proper time interval?
O-7 Combustion safety controls
    a. Does stack switch or cad cell bring on unit within proper time interval after call for heat?
    b. Does cad cell "see" flame properly during firing?
O-8 Primary control
    a. Is there power at T terminals?
    b. Does switch respond to signals from combustion safety device?

TABLE 21–4 (continued)
Troubleshooting Procedures for Oil-Fired Furnaces

Control Components (continued)

    c. Does safety switch open on call from combustion safety?

Burner and heating components

O–9 Oil supply
    a. Is there oil in the tank?
    b. Has sludge plugged the intake?
    c. Has water gotten into lines?
    d. Are all line fittings tight?

O–10 Piping
    a. Check oil lines for oil or air leaks

O–11 Pump
    a. Is pump running on a call for heat?
    b. Is pump noisy?
    c. Is pump producing the correct oil pressure?

O–12 Nozzle
    a. Is nozzle proper size?
    b. Is nozzle clean?
    c. Is spray pattern correct?

O–13 Ignition transformer
    a. Does transformer produce proper voltage?
    b. Is electric connection good from transformer to electrodes?

O–14 Ignition electrodes
    a. Check insulators to make sure they are clean and not cracked
    b. Are electrodes properly located and spaced?
    c. Are points proper shape?

O–15 Combustion blower
    a. Does combustion blower come on on a call for heat?
    b. Check blower adjustment

O–16 Fire pot
    a. Is fire pot lining intact?

Burner and heating components (continued)

O–17 Heat exchanger
    a. Check for cracks or burn-out
    b. Is there any flame impingement?
    c. Is inside of exchanger clean?

O–18 Barometric damper
    a. Adjust damper for proper draft over fire

Air-handling components

19. Blower
    a. Is blower wheel clean?
    b. Check for correct rotation
    c. Check speed of rpm
    d. Is wheel noisy?

20. Blower motor
    a. Does motor operate when blower control closes?
    b. Is motor noisy?
    c. Does motor run hot?

21. Belt and drives
    a. Is belt broken or cracked?
    b. Check belt tension
    c. Are pulley sizes correct for cfm required?

22. Filter
    a. Is filter in place?
    b. Does filter need cleaning or replacing?
    c. Is filter efficiency proper for job?

23. Outside air
    a. Check for proper quantity of outside air

24. Ductwork
    a. Is ductwork leaky?
    b. Is ductwork too small?
    c. Does duct run in uninsulated space?

TABLE 21–5
Troubleshooting Procedured for Electric Furnaces

Electric power components
1. Disconnect
    a. Check main disconnect for power
    b. Check unit disconnect for power (may be circuit breakers)
2. Fuses (circuit breakers)
    a. Check main fuses for opens
    b. Check unit fuses for opens (may be circuit breakers)
3. Control transformer
    a. Check for voltage on primary side
    b. Check secondary voltage
    c. Check for 60 cycle hum

Control components
E–4 Thermostat
    a. Check setting
    b. Check for power to thermostat from transformer
    c. Do contacts close on call?
    d. Check location (sun, drafts, etc.)
    e. Is thermostat level?
E–5 Limit control
    a. Is limit control closed in normal operating conditions?
    b. Is thermal link intact in power circuit?
E–6 Blower control
    a. Is blower control normally open, and does it close on temperature rise?
E–7 Blower relay
    a. Are relay contacts open when coil is not energized?
    b. Do contacts close when coil is energized?
E–8 Sequencer
    a. Does sequencer start staging on a call for heat?
    b. Does each stage come on in proper sequence?

Control components (continued)
    c. Does sequencer step each stage off in proper sequence?
    d. Does sequencer signal blower relay?

Heating components
E–9 Wiring
    a. Are all connections tight?
    b. Is insulation intact?
E–10 Elements
    a. Do elements heat up when energized?
    b. Are all insulators in place?
    c. Are elements sagging excessively?

Air-handling components
19. Blower
    a. Is blower wheel clean?
    b. Check for correct rotation
    c. Check speed of rpm
    d. Is wheel noisy?
20. Blower motor
    a. Does motor operate when blower control closes?
    b. Is motor noisy?
    c. Does motor run hot?
21. Belt and drives
    a. Is belt broken or cracked?
    b. Check belt tension
    c. Are pulley sizes correct for cfm required?
22. Filter
    a. Is filter in place?
    b. Does filter need cleaning or replacing?
    c. Is filter efficiency proper for job?
23. Outside air
    a. Check for proper quantity of outside air
24. Ductwork
    a. Is ductwork leaky?
    b. Is ductwork too small?
    c. Does duct run in uninsulated space?

Checking Individual Parts

After determining the component in a heating system that is not functioning correctly, a troubleshooter must identify which part, or parts, in that component is working incorrectly. To do this the troubleshooter starts examining the individual parts for proper operation. This procedure is simplified by using troubleshooting charts.

21.4 TROUBLESHOOTING CHARTS

The components and parts table, troubleshooting tables, and procedure tables shown in this text are all used together in troubleshooting a heating system. On the components and parts tables the main components and parts for the various types of furnaces are listed and keyed to the troubleshooting and procedural tables. The troubleshooting tables show the various types of furnaces keyed to types of complaints. The troubleshooting procedure chart shows the heating unit parts related to each unit component, and these are identified by number and listed below each component. One or more checks to make concerning each part are also listed and are identified by letters.

Tables 21-3, 21-4, and 21-5 are troubleshooting procedure charts for gas-fired furnaces, oil-fired furnaces, and electric furnaces. The tables contain the list of parts and descriptions of the procedures for checking each part.

Each troubleshooting procedure chart contains four lists. The four lists are for the main components: electric power, controls, burner and heating components, and air-handling components. The electric power component list and the air-handling list are basically the same for all three types of furnaces, but the controls and heating components lists are different for each.

Checking Unit Parts

The number in the columns under the complaints on the troubleshooting chart indicate the order in which the parts should be checked for faults. For instance, if it is reported that the temperature in a building using a gas-fired furnace fluctuates from the setpoint temperature on the thermostat, then refer to that portion of Table 21-2 that shows the steps to take in troubleshooting a gas-fired unit.

In the column under the heading, "heat fluctuates," the number 1 is found across from "thermostat" in the components column. The thermostat is the first part to check. If the thermostat is operating correctly, the troubleshooter goes to number 2 and the chart, found across from "blower control." The troubleshooter checks the operation of the blower control. The procedure to follow in checking the parts of the gas-fired furnace is shown in Table 21-3 and is also described later in this chapter. If the blower control is operating correctly, then the troubleshooter continues to number 3 and checks the part indicated there. This procedure is followed until the faulty part is found.

EXAMPLE 21-1

A complaint is received that a building heated by an oil-fired furnace is not heating properly (insufficient heat).

Solution

Under the section for oil-fired furnaces in the troubleshooting chart in Table 21-2, the troubleshooter locates the complaint heading "Insufficient heat."

The first part of the unit to check is the thermostat. Previous investigation would have indicated this problem to be in the controls section. Using the troubleshooting procedure chart for an oil-fired furnace (Table 21-4), under the section for control components, the first check for the thermostat is to make sure that the thermostat is set higher than the room temperature (in other words, to see whether it is calling for heat). If the thermostat is calling for heat and the furnace still is not on, then the troubleshooter should check whether there is electric power to the thermostat from the transformer. If that check does not find the problem, the troubleshooter should see whether the thermostat contacts are closing when they should. The troubleshooter proceeds in this manner until the fault is found.

If the fault is not found in the thermostat or the electric control circuit, then the troubleshooter will see from Table 21-2 that the second part to check under "Insufficient heat" is the limit controls. The limit controls could be short-cycling the unit, preventing it from producing its full rated capacity. Using the troubleshooting procedure (Table 21-4), the troubleshooter locates the procedure for checking the limit control under the control components section. The troubleshooter follows this general procedure until he or she identifies the faulty part.

Identifying Unit Parts.   Table 21-2 is a troubleshooting chart for gas-fired furnaces, oil-fired furnaces, and electric furnaces. The major components are listed on the left side of the table, with the related parts listed below each component. The most common complaints are shown at the top right of the table. In each column below the listed complaint, there are numbers across from the various parts. These numbers start at 1 and continue in order. The numbers give the order in which the parts should be checked.

21.5 SUMMARY

Like all other mechanical equipment, heating equipment needs maintenance to keep it working properly. Occasionally it needs service to adjust or repair malfunctioning parts. The process of investigating the problem when a unit stops working correctly is called troubleshooting. Troubleshooting takes place after the equipment is no longer doing what it is supposed to, and before the actual service work is preformed to correct the problems.

21.6 TERMS TO KNOW

troubleshooting
service
malfunction

CHAPTER
22
Troubleshooting Individual Parts

22.1 INTRODUCTION

The positive identification of faulty parts is perhaps the most important part of any troubleshooting process. Certain procedures, often involving the use of test instruments, and generally involving observation and testing, help to identify those parts.

The description of some of the main parts and controls in earlier chapters included some troubleshooting and service procedures. This section covers some specific troubleshooting procedures that can be used for individual parts. Many of the procedures listed on the troubleshooting procedure charts are obvious. They do not need explanation and will not be covered in this text.

22.2 ELECTRICAL CONTROL COMPONENTS

Basic electrical power components are common to almost all types of furnaces. These components are disconnects, fuses, and (in some cases transformers). Procedures for troubleshooting each of these follows. Some operating and safety controls also are common to most furnaces. Among them are thermostats, limit controls, blower controls, and blower relays.

Disconnects

First, a troubleshooter should make a visual check of the main electrical disconnect to see if the switch is closed. If it is closed, but the unit does not come on, then a voltmeter should be used to check for power. The check is made first on the line side and then on the load side of the disconnect. The voltmeter is set at a higher voltage setting than the expected voltage for the system, to protect the meter against damage due to overload. Then one probe of the voltmeter is placed on each of the two terminals on the line side of the disconnect. Figure 22-1 illustrates this procedure.

For a three-wire system, the three legs of the system will have to be checked in combinations of two. Use of the voltmeter on the line side of the disconnect will indicate if there is power to the disconnect. If there is no power on the line side, the electric utility that provides the power

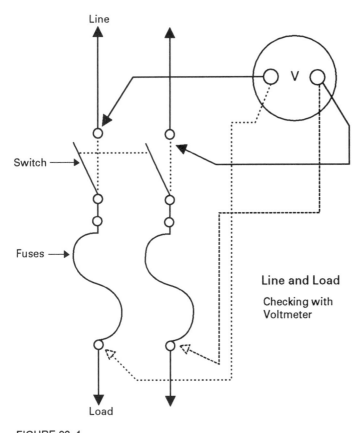

Line

Switch ⟶

Fuses ⟶

V

Line and Load

Checking with
Voltmeter

Load

FIGURE 22–1 _____
Voltage on the line side and load side of an electrical disconnect are
checked with a voltmeter

should be notified. After checking the voltage on the line
side the same thing should be done on the load side of the
disconnect. If there is power on the load side of the main
disconnect, but the unit still does not run, then the unit
disconnect should be checked, using the same procedure
as that outlined for the main disconnect.

Fuses and Circuit Breakers

If there is power on the line side of a fused disconnect or a
circuit breaker, but not on the load side, the fuses should
be checked. Two methods can be used: (1) the fuses can be
removed and checked for continuity with an ohmmeter, or
(2) they can be checked in place with a voltmeter.

FIGURE 22-2
A cartridge fuse can be checked for continuity to see if it is burned out

To check with an ohmmeter, one of the meter probes is held to each end of the fuse, and the ohm value of the fuse is read on the meter. If the fuse is good, the meter will show very little resistance. If the fuse is burned out, the reading should show infinite resistance. Figure 22-2 shows the proper procedure for checking a fuse with an ohmmeter.

To check fuses in place, the troubleshooter places the voltmeter probes on the line side of one fuse and the load side of another. A volt reading on the meter indicates that the fuse with the probe on the load side is good. No reading indicates that fuse is burned out. The procedure should be reversed for checking the other fuse(s) in the system. See the procedure for checking fuses with a voltmeter in Figure 22-3.

Some circuit breakers have an indicating switch built in with their disconnect switch. If the circuit breaker has opened, the toggle switch handle will have moved to the off position, and a red signal will indicate it is open.

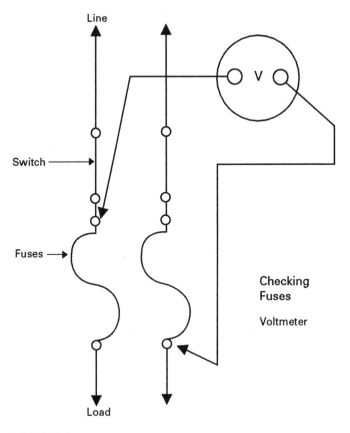

FIGURE 22–3
Fuses can be checked while in place by using a voltmeter and cross-checking them

Transformer

If there is electric power on the load side of the unit disconnect and all fuses are good, but the unit still will not run, the control transformer should be checked. A voltmeter can be used to take readings on the primary and secondary sides of the transformer. Figure 22-4 shows methods used for checking a transformer. The voltage on the primary side should be the same as that on the load side of the unit disconnect. The voltage on the secondary side should be within plus or minus 10 percent of the rated secondary voltage of the transformer.

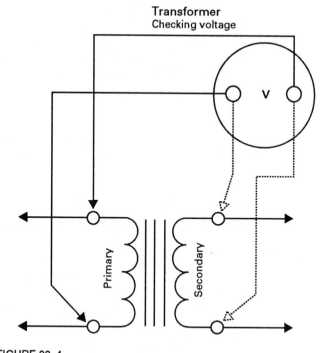

FIGURE 22–4 _____

A transformer can be checked with a voltmeter on both the primary and secondary side

22.3 OPERATING AND SAFETY CONTROLS

Operating and safety controls are those controls that control the operation of the individual parts in a heating unit, and protect the parts during operation to prevent damage to the equipment or operators.

Thermostats

The first check to make of a thermostat is visual. Is the setpoint temperature higher than the room temperature? If it is, check to see if the thermostat contacts are closing by placing a jumper wire across the terminals. If the unit comes on with the jumper wire in place the thermostat is faulty and should be replaced. See the procedure in Figure 22-5.

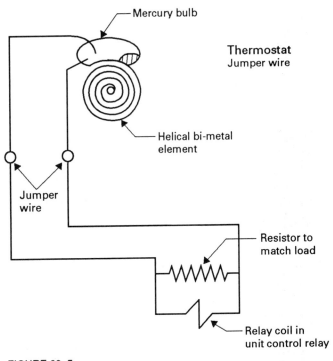

FIGURE 22–5
An electrical jumper wire across the terminals of a thermostat show whether the contacts in the thermostat are open

A thermostat should always be located on an inside wall where it will sense average room temperature. It should not be positioned where drafts from doors or windows will affect it, or where direct sunshine will strike it. Any one of these locations can cause a reading by the thermostat that is very different from that of the room air. The thermostat should always be mounted level on the wall so the contacts function properly.

A thermostat **anticipator** has to be set to coincide with the amperage draw of the gas valve and any other electrical loads on the thermostat circuit. The amperage in the thermostat circuit is checked with an ammeter while the unit is in operation. To check the amperage on the circuit a n ammeter with a needle indicating amperage can be used, but it is usual to increase the amperage reading by inserting a coil of control wire with 10 turns into the control circuit, and then the jaws of an ammeter are placed through the loop of wire. See Figure 22-6 for the wiring arrangement. The 10 turns of wire multiply the ammeter reading

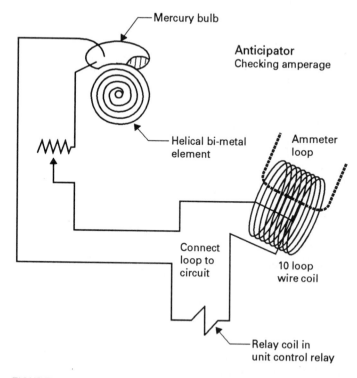

Mercury bulb

Anticipator
Checking amperage

Helical bi-metal
element

Ammeter
loop

Connect
loop to
circuit

10 loop
wire coil

Relay coil in
unit control relay

FIGURE 22–6

Current draw in a control circuit can be checked with a standard ammeter by use of a multiplying loop in the control wiring circuit

by 10. With the thermostat calling for heat and the furnace
running, the amperage is read on the ammeter. This reading is divided by 10. The result is the amp draw on the circuit. The anticipator is set at this reading by adjusting the
anticipator lever in the thermostat.

Most digital ammeters can read low enough amperage
values so it is not necessary to use the multiplier technique
in using them on an anticipator circuit.

Limit Controls

To check a limit control, a troubleshooter disconnects the
blower from the power by removing one of the wires from
the blower motor or blower control. A high temperature
thermometer is placed in the supply plenum of the furnace. The furnace is then started by turning the thermostat
to a setting higher than the room temperature. The limit
control should open and turn the unit burner off when the
thermometer in the plenum indicates the temperature that

the control is rated for, usually about 180°F to 200°F. The blower should then be started by reconnecting the wire. When the plenum temperature drops below the rated temperature for the control, the unit should come back on.

A limit control can also be checked by disconnecting the motor leads, and applying heat to the sensing element while checking continuity across the contacts with an ohmmeter. The contacts should close when the sensing element closses the contacts. See Figure 22-7.

Blower Controls

A blower control is an electrical control used to bring a furnace blower on, either when a furnace is first fired, or after a short delay to allow the unit to heat up. Proper operation of a blower control can be checked by observation or checked manually.

To check a blower control by observation, set the unit thermostat to call for heat, after the heating unit comes on, the blower should come on as soon as the air going through the heat exchanger reaches the setpoint temperature of the blower control. The temperature of the air can be checked by a thermometer in the supply plenum. The setpoint temperature is adjustable on some blower controls and is preset on others. If the blower turns on when

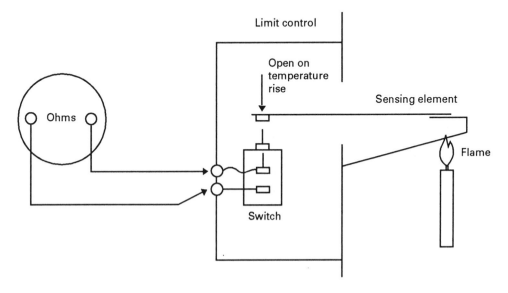

FIGURE 22-7
A limit control can be checked for proper operation by applying heat to the sensing element while monitoring the opening and closing of the contacts

the air temperature in the supply plenum reaches the set-point temperature of the blower control, the control is functioning properly.

The setpoint temperature for a blower control usually is somewhere between the room thermostat setpoint and the limit control setpoint. If the setpoint is less than room temperature setpoint the blower will run before the furnace comes on, and if it is set above the limit control setpoint the blower will never come on.

To check the blower control on and off cycle, the blower control should open and turn off the blower after a delay of approximately 30 seconds to 1 minute after the heating unit shuts off.

To check a blower control manually, disconnect the control from the blower motor circuit, and remove the control from the furnace. Set the control in a stand so heat can be applied to the sensing element by an outside source, such as a torch. Light the torch and place the leads of an ohmmeter across the terminal connections of the blower control. This procedure is shown in Figure 22-8.

With the lighted torch heating the sensing element observe the ohmmeter. It should be reading infinity at first, and should show continuity when the heating element is heated to the set point temperature, indicating that

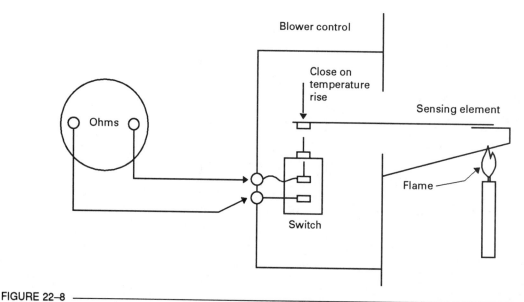

FIGURE 22–8

Blower control operation can be checked by applying heat to the sensing element while monitoring the opening and closing of the contacts the switch is closed. If the switch in the control does not close, showing continuity on the meter, the control should be replaced.

Sure-Start Switches

A sure-start switch is a blower control that is actuated by both temperature and/or by a low voltage signal from the thermostat. Sure-start switches often are used as blower controls on heating units in which the air pattern is not naturally established by gravity on start up. Such heating units as downflow, horizontal, and rooftop units often use sure-start switches as blower controls.

A sure-start switch can be checked for proper operation by using the same procedures described herein for a blower control, plus it should be checked for operation when the control voltage leads are energized. To check a sure-start switch for proper operation on temperature rise, the operation of the switch is checked to see if it closes and turns on the heating unit blower at a predetermined air temperature. The low voltage control wiring from the thermostat is disconnected from the switch. A high temperature thermometer is placed in the supply air plenum of the furnace being checked, and the thermostat is then set to call for heat. The heating unit blower should turn on when the supply air temperature reaches the control setpoint, and when the room thermostat is satisfied the blower should turn off when the supply air temperature drops below the sure-start switch setpoint.

To check the operation of a sure-start switch by the heating unit thermostat control circuit, remove the control from the furnace, but keep the wiring from the thermostat on the control. Set the thermostat to call for heat, and check the voltage across the sure-start terminals across the leads. If the control circuit is sending the proper signal to the sure-start switch, regular control voltage will be registered. See this procedure in Figure 22-9.

To check the heater in the sure-start switch, disconnect the switch from the control circuit by removing one of the thermostat leads at the switch, and check across the heater by placing the leads from an ohmmeter on the control terminals. This procedure is shown in Figure 22-10. A resistance reading across the coil should be seen. If the ohmmeter indicates no resistance, the heater is open, and the sure-start control should be replaced.

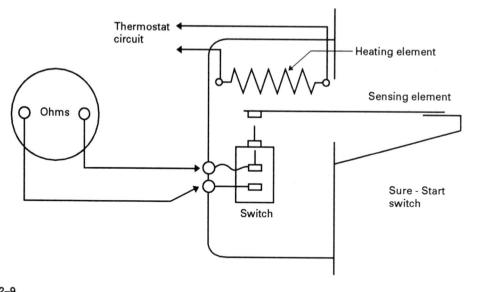

FIGURE 22–9

To check the operation of a sure-start switch when voltage is applied to the heating element the function of the contacts should be observed

FIGURE 22–10

To check the function of a sure-start switch by temperature, apply heat to the sensing element while monitoring the opening or closing of the contacts

Blower Relays

If the thermostat in a heating unit has a blower control switch on it, there will be a relay in the control system operated by that switch for continuous blower operation. The coil of the relay is wired in the control circuit in series with the blower control switch in the thermostat. The contacts in the relay are wired in parallel with the contacts in the blower control. When the blower control switch on the thermostat is set at "continuous," the relay brings the blower on. With the contacts in this relay wired in parallel with the contacts in the blower control either one or the other can operate the blower motor. See a blower relay wired into a control circuit in Figure 22-11.

A relay should be checked for proper operation by checking the coil for continuity with an ohmmeter. See Figures 22-11, 22-12.

The contacts are checked, in both the open and closed positions, by disconnecting one of the wires running to the contacts and checking across the contacts with an ohmmeter. See Figure 22-13 for this procedure.

The check should be made with the relay coil energized, and again with it deenergized.

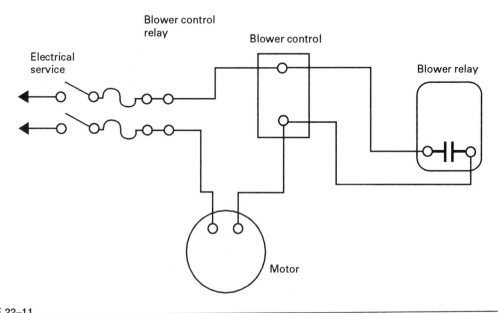

FIGURE 22-11
When auxiliary control of a blower motor is desired, a blower relay is employed

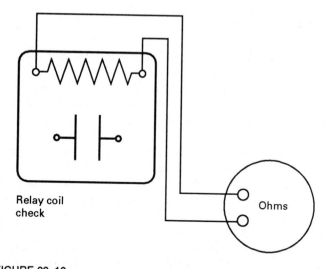

FIGURE 22-12 _____
To see if a relay coil is good an ohmmeter can be used to check across
the coil

Time-Delay Relays

Some heating units use a form of a time delay relay for energizing the blower motor. Rooftop units and many larger heating units are among these. To check a time delay relay

Relay contacts
Checking

FIGURE 22-13 _____
To check the operation of a relay, an ohmmeter is used to check for continuity across the contacts while control voltage is applied to the coil

operation, the relay should be energized by setting the heating unit thermostat to call for heat. Voltage across the time delay relay's terminals can be checked with a voltmeter. With the unit thermostat calling for heat, the voltage should be regular control voltage. See Figure 22-14 for the procedure to use. With the relay energized as above, the unit blower should be observed. It should turn on within a reasonable, 40- to 60-second, interval. To check the relay for off time, the relay should be deenergized and the blower observed. The blower should either turn off immediately or after a reasonable, 40- to 60-second, time delay.

22.4 GAS FURNACE PARTS

The two general types of gas furnace burners are described in Chapter 7. This section describes troubleshooting procedures to use for some of the components used with the burners: the gas valve, thermocouple, pilot safety, and pilot burner.

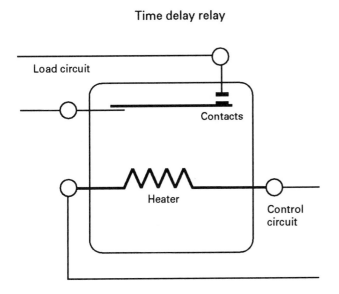

Time delay relay

Load circuit

Contacts

Heater

Control circuit

FIGURE 22–14
A time-delay relay functions when a heater in the relay is energized by control voltage

Gas Valve

Operation of a gas valve can be checked by using a voltmeter to see if there is power to the valve on a call for heat. The voltmeter should be set to the appropriate voltage scale and one probe placed on each terminal of the solenoid on the valve. The thermostat is set to call for heat. Control voltage should show on the voltmeter. If there is power to the terminals, the gas valve should open. This is shown in Figure 22-15. The opening of the valve can be checked by observing the burners. If the burner does not come on, the combustion safety controls should be checked before the valve is condemned.

If the combustion safety controls are functioning prop-

Gas valve - solenoid check

Control Circuit

Volts

Thermocouple lead

Valve handle

Pilot safety coil

Pilot solenoid

Pilot gas line

Pilot valve

Main gas line

To burner

Main valve

Pressure regulator

FIGURE 22-15

A voltmeter is used to check whether electricity is reaching the gas valve when a gas fired unit fails to come on during a call for heat erly and the burner still does not come on, the troubleshooter should disconnect the control wiring from the solenoid and check the solenoid for continuity with an ohmmeter. An ohmmeter reading indicating an open circuit on the coil shows that the coil is burned out. If the coil is all right and the valve still does not open, the problem is within the valve itself. The valve should then be replaced. See Figure 22-16.

If a gas valve sticks open and the burner will not shut off after the thermostat is satisfied, the gas should be shut off at the unit shutoff valve, and the valve should be changed.

FIGURE 22-16
The solenoid in a gas valve can be checked for continuity by using an ohmmeter to check across the terminals while the valve is disconnected from the control system

Thermocouple

If on lighting a pilot light, the pilot stays lighted while the pilot switch is held on the valve, but goes out as soon as the switch is released, the thermocouple should be checked. To check this part, the connecting lead to the pilot safety valve, or the gas valve if the safety is in it, is disconnected. With the pilot flame burning, a reading is taken with a milivoltmeter between the terminal on the end of the thermocouple lead and the outer part of the lead cable. The positive lead probe of the milivoltmeter is placed on the outside conductor and the negative lead probe on the inside lead or terminal. The procedure is shown in Figure 22-17. The acceptable minimum voltage for most valves is approximately 18 milivolts. (Check the proper data in the valve manufacturer's data sheets for the valve used.) If the reading is less than the minimum allowed for the specific valve, the thermocouple should be replaced.

Pilot Safety.   If the thermocouple checks out all right but the pilot still goes out as soon as the pilot reset button is released, the pilot safety is faulty and should be replaced.

FIGURE 22–17

A thermocouple is checked for proper operation by applying heat to the couple while checking the millivolt on the leads

Pilot Burner. The pilot burner is checked first by visual inspection of the flame. The flame should be steady, with about the last one-third of the flame surrounding the tip of the thermocouple. If the flame is too low, the pilot gas valve should be adjusted, the screen in the pilot burner should be cleaned, or the small hole in the pilot orifice should be cleaned. If the flame is too high or is lifting off the burner, the pilot gas input should be reduced by turning down the pilot gas valve or pilot adjustment screw.

Electric Ignition. The components of a so-called electric ignition system include both electric and electronic parts. Normally a special gas valve is used, with spark or hot surface ignition, and an electronic control panel which contains printed circuits with electronic control devices and relays. A compatible electronic thermostat also is included in most electric/electronic control systems.

Troubleshooting and servicing of electronic control systems normally is limited to tests that are made only on the exterior of the control circuit terminals, and with special meters that are calibrated to work properly on the devices. Care must be taken to not short circuit certain circuits in the control system, or permanent damage will occur to the electronic devices.

The control devices manufacturer's instructions should always be followed when checking an electronic control system.

22.5 OIL FURNACE PARTS

Oil furnace parts are described in detail in Chapter 8. Procedures for checking the separate parts of a burner and the combustion safety control specifically related to oil furnaces are covered here.

Oil Burner

If oil fails to get to the burner, the following should be checked: the oil supply in the tank, piping for leaks, fuel pump, and/or the fuel pump motor. A dipstick can be used to check the oil supply. If the tank has been allowed to run dry and then filled, sludge may be present in the line or pump. If the pump whines when it comes on, the suction pressure should be checked with a gauge on the intake side of the pump. The suction pressure should be less than 5 in. if the system is gravity feed, or not more than 10 in. if the tank is below the level of the burner.

If the vacuum is all right, either the bleed valve should be opened, or the oil line disconnected from the pump to the gun assembly while the pump is running. Oil should come out of the disconnected line. If oil does not, the line is plugged, there is a restriction in the line from the tank to the pump, or the pump screen is plugged. To check for a plugged or restricted line, the troubleshooter should disconnect the line at the burner pump intake and at the oil tank. Air pressure should be used to force the oil out of the line and to clear any obstructions. If air will not go through the line, the line will have to be replaced. To check for a plugged screen, remove the head from the pump and check the screen.

A faulty burner motor is checked by following the same procedure as used for any other electric motor. If the fuel pump is faulty, the problem usually will show up by causing excessive amp draw on the pump motor. Mechanical failure of the pump will make the motor turn hard, overloading the motor.

Oil Nozzle

A poor or unbalanced flame can indicate a faulty nozzle. To check for this problem, the troubleshooter can inspect the flame with an inspection mirror. An inspection mirror is a small metal mirror on a long handle. The mirror can be inserted into the combustion chamber through the inspection door. It is used to inspect the inside of the combustion chamber or the burner face while the furnace is being fired. A visual inspection of the nozzle can be made by removing the entire burner assembly from the blast tube, and removing the nozzle from the oil pipe. The strainer on the inlet side of the nozzle also should be checked for dirt. If any of these problems are found, the nozzle should be replaced with a new nozzle of the same type and size.

Ignition Transformer

A faulty ignition transformer can cause a weak ignition spark, resulting in delayed or intermittent ignition. To check the transformer, the troubleshooter should first check the primary voltage and make sure it is within 15 percent of the rated voltage for the unit. If it is not, excessive voltage drop to the unit should be considered. Figure 22-18 shows the proper way to check the ignition transformer with a voltmeter. If the primary voltage is all right,

Ignition transformer

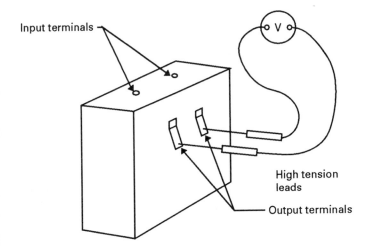

Ignition transformer

FIGURE 22–18 _____
Voltage input and output on an ignition transformer can be checked with a voltmeter. High tension leads should always be used on the output

then the secondary voltage output should be checked. It should be within the limits given by the manufacturer for the unit. Usually the secondary voltage is approximately 10,000 V from either terminal to ground. Be sure to use high voltage probes when checking on the secondary side of an ignition transformer. If the transformer is faulty, it must be replaced.

Ignition Electrodes

Electrodes that are burned on the end, or misaligned in the unit, can be the cause of faulty ignition. The electrodes should be checked visually. If the ends are burned, they can be dressed with a file to a near point with a blunt end. They should then be replaced in the unit as the manufacturer of the unit recommends. Normally this recommendation is for the tips 1/8 in. beyond the nozzle. The tips are normally 1/2 in. above the nozzle. Figure 22-19 shows the proper alignment of ignition electrodes. In all cases, the exact alignment should be as specified by the burner manufacturer.

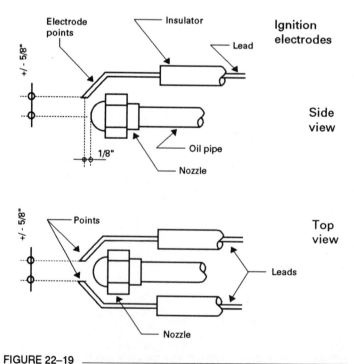

FIGURE 22–19

Proper adjustment of the ignition electrodes is important for firing of an oil fired furnace

Combustion Chamber and Heat Exchanger

A visual check of the combustion chamber and heat exchanger should be made with an inspection mirror to determine if there are any breaks in the lining of the chamber or cracks or burnouts in the exchanger. The burner can be removed from the burner port for examining. Some units have inspection plates that can be removed from the upper part of the exchanger to facilitate cleaning and inspection.

Barometric Damper. A **barometric damper** is a tee installed in the vent of a furnace. The open side of the tee faces the furnace room. This side is covered with a damper mounted on pivots. The damper is balanced so it will open and allow room air to enter the vent if the draft in the vent exceeds a predetermined limit. The barometric damper is used on heating units with power burners. See Figure 22-20.

The barometric damper should be adjusted to provide the proper draft over the fire. The draft is necessary for

Barometric damper*

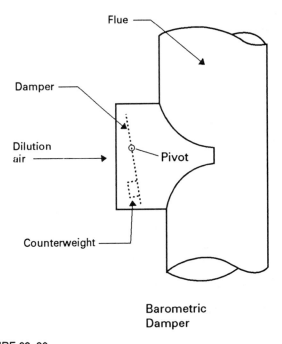

Barometric
Damper

FIGURE 22–20

A barometric damper is used to prevent draft from lifting the flames out of a combustion chamber proper combustion and to prevent the flame from being pulled up out of the combustion chamber.

Draft over the fire is checked by using a draft gauge or an inclined manometer. The reading is taken with the unit operating. The gauge is inserted in the small hole in the inspection door in the front of the furnace. The draft over the fire should be between 0.02 in. wg and 0.04 in. wg. Another reading should be taken between the flue outlet and the barometric damper. This reading should be between 0.04 in. wg and 0.06 in wg. There should be no more than 0.02 in. wg difference between the reading at the flue outlet and the inspection door. If the readings indicate more than 0.02 in. wg difference, the heat exchanger flue passages may be clogged with soot. Too low a difference indicates air may be leaking into the fire areas through cracks in the unit. See Figure 22-21 for an illustration of the procedure.

Combustion Safety Control

On an oil-fired unit a troubleshooter should check the fol-

FIGURE 22–21

The draft through a combustion chamber is measured by use of a draft gauge lowing parts of the combustion safety control: stack switch, cad cell, and cad cell detector.

Stack Switch.   To check a stack switch for proper operation, the control should be placed "in step" by pushing the reset button or setting the lever. If the burner starts, the unit was just off on safety shutoff. If the burner fails to start, then place a jumper wire across the cold contact terminals on the detector. This procedure is shown in Figure 22-22. If the burner fails to start with the jumper wire in place, the voltage between terminals 1 and 2 on the control should be checked. If there is less than 105 V across them, the voltage across the burner terminals should be checked. These are the terminals to which the burner leads are attached. The troubleshooter should make sure the thermostat is calling for heat and that the cold contacts are made. If, when the reset button is pushed, there is no voltage, then the fault is in the primary control. The problem could be dirty burner relay contacts. Contacts can be

FIGURE 22–22 _____
Proper operation of a stack switch can be checked by using a jumper
wire to by-pass the control

cleaned with a contact cleaner. If the internal circuitry is faulty, the control should be replaced.

Cad Cell. A troubleshooter checking a cad cell safety control on a safety shutoff should make sure the thermostat is calling for heat and then reset the safety switch lever on the primary control. If the burner does not start, the flame detector wires on the primary control should be disconnected. If the burner starts, the fault is in the flame detector circuit. The problem could be shorted leads, a flame detector exposed to light, or a short circuit in the detector. Each of these possibilities needs checking. Faulty parts must be repaired or replaced. If the burner still does not start, the voltage between the black and white leads should be checked. No voltage or low voltage indicates no power to the control. The power should be checked. The limit control could be open or faulty.

Cad Cell Detector. To check a cad cell detector for proper operation, the troubleshooter should disconnect the cad cell from the flame detector terminals on the primary control. A jumper wire should be placed across the terminals. The burner should be allowed to run with the jumper in place. With an ohmmeter, the troubleshooter should check the resistance across the detector terminals. See Figure 22-23. If the resistance exceeds the recommended reading for the particular type of cell (approximately 1500 ohms), then the cell is dirty or faulty; the wiring to the cell is faulty; or the cell is crooked in its mounting and is not seeing the flame directly. If the burner locks out on safety while firing with the jumper in place, the fault is in the primary control.

22.6 ELECTRIC FURNACE PARTS

The two parts of an electric furnace not covered elsewhere in this book are the elements and the sequencer. This section gives some procedures that can be used for checking these parts.

Electric Heating Elements

A complaint of not enough heat can result if some of the elements in an electric furnace are burned out. Because

Light bulb

Cad cell

Cad cell
check

Ohms

FIGURE 22–23
A cad cell can be checked for proper operation by use of a light and ohm-meter

there are multiple elements in an electric furnace and they are brought on in sequence, burned-out elements often are not discovered until the outside temperature drops low enough to need all of them for heat. There are two ways to check for burned out elements. The first is to check the amperage draw through each element circuit when the thermostat is calling for heat. If an element circuit shows no amperage draw, then it is burned out. The second check is to disconnect each element from the power and use an ohmmeter for checking continuity A burned-out element will give an infinite ohm reading. The procedure is shown in Figure 22-24.

Sequencer

To check a sequencer for proper operation, the troubleshooter should make sure that the thermostat is calling for heat. Then she or he should use an ohmmeter to check the on time for each set of contacts. There should be about a 30 to 50 sec delay between each set. In some relay systems, if any one set of contacts fails to close, the others following it in the normal sequence also will fail. If any contacts fail to close or are welded shut, the sequencer should

FIGURE 22–24
Voltage across the heating elements of an electric furnace can be checked to verify operation of the sequencer

be changed. The procedure for checking a sequencer is shown in Figure 22-25.

22.7 AIR DISTRIBUTION PARTS

Another section of a system that is generally the same for all units is the air distribution system, which includes the blower, filter, and ductwork. Procedures for checking each of these are discussed here.

Blower.   To check the blower for proper operation, the troubleshooter should make a visual check to see if the pulleys are aligned so the belt runs straight. To check the alignment of the pulleys, a straight rod or a dowel can be laid in the pulley grooves, from one pulley to the other, as shown in Figure 22-26. The grooves in the pulleys should line up with the rod and with each other. If they do not line up, one or the other of the pulleys should be moved

Electrical sequencer diagram

Operating coil

Ohms

Element one

Time delay coil element two

Time delay coil element three

Element two

Electrical Sequencer

Checking contacts

Element three

FIGURE 22-25 _____
Operation of the contacts in a sequencer can be checked with an ohmmeter with the elements disconnected

on its shaft so they do line up. The setscrew that holds the pulley in place on the shaft should be loosened and the pulley moved so it will line up with the other one.

Tension on the blower belt should be checked to determine that the belt is neither too tight nor too loose. If it is too loose, the belt will slip on startup. The correct tension should

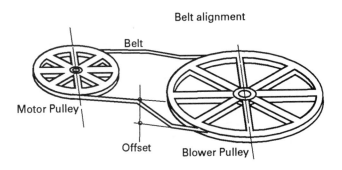

Belt alignment

Belt

Motor Pulley

Offset

Blower Pulley

FIGURE 22-26 _____
A blower belt must be properly aligned on both motor pulley and blower pulley

allow about 1 in. of play when the two sides of the belt are pulled toward each other. This procedure is shown in Figure 22-27. If the tension is not right, it can be adjusted by loosening the motor on its mounts and tipping it either toward the blower or away, as needed. The motor mount should have provisions for this adjustment and an adjusting screw for setting the motor to the correct position.

The blower bearings should be checked for wear at the same time by manually moving the shaft in the bearings and observing the play. If the play is excessive, the bearing should be changed.

The blower wheel should always be checked for proper direction of rotation.

Blower Motor.   The blower motor is checked for worn bearings in the same way that the blower was. The blower motor should also be checked for overheating by placing a hand on it while it is operating. If the motor is too hot to keep a hand on it for any length of time, it is overloaded. The amperage should be checked on the motor circuit and this compared to the rated amperage on the nameplate. If the operating amperage exceeds the rated amperage, then the motor should be replaced with one of a higher horsepower rating.

It may be difficult to check blower motor operation without removing the blower vestibule panel. If the blow-

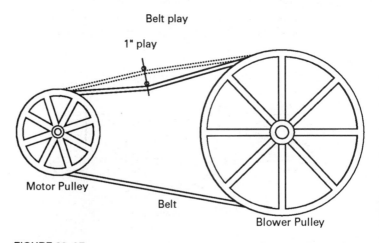

FIGURE 22–27

When a blower belt is properly tensioned play of about 1 inch is found on one side of the belt er is operated with this panel removed, the motor may overload and burn out. Care should be taken to keep the blower panel in place as much as possible, and to not run the motor longer than necessary with the panel open.

Filters. Mechanical filters should be checked visually on a regular basis. If the surface coating of dust and dirt particles starts to close off the air passage, the filters need to be cleaned or changed. If the filters are the replaceable type, they usually can be cleaned carefully perhaps once; then they will need to be replaced. If the filters are permanent, they can be washed in water and detergent and replaced in the unit.

Electronic cleaners usually have a test button or meter on the front of them that shows whether they are working. They should be checked occasionally. The prefilter should be pulled out and cleaned regularly. The electronic cell will need cleaning only about twice a season in a normally dirty atmosphere. If the cell is used in a building where the air is very dirty, the cell should be cleaned more often. Service on the unit should be performed according to the manufacturer's instructions.

Ductwork.   Problems with the ductwork may show up when the system is balanced. The procedure used for balancing a system is found in Chapter 18.

22.8 FOLLOW-UP

After identifying the faulty part(s) of a systems, the next step is to correct the problems or replace faulty parts. Usually the troubleshooter does this work, but in some cases a service specialist may be needed, especially if a major part or control must be repaired or replaced. The repair or replacement of the parts should be made. Then the unit should be thoroughly tested to make sure that it operates properly.

In most cases, replacing the parts rather than repairing them is less expensive and better in the long run. Most service people are not equipped to repair parts and controls. Also, the cost of the parts usually is much less than the cost for labor to effect a repair. New parts normally carry a warranty, while repaired parts do not, and the charge for repeat failure is less when new parts are used.

22.9 SUMMARY

A systematic investigation is made in troubleshooting a piece of equipment. First, the troubleshooter must understand the operation of the equipment and the system. Second, the troubleshooter must understand what the complaint is and its probable cause. Third, the troubleshooter must decide what general part of the unit is affected by the breakdown. Fourth, the troubleshooter must check that section of the system to identify the actual part (or parts) that is faulty.

After the part that is not functioning correctly is identified, it must be adjusted or replaced as necessary to make the unit functional again. After the repair, the unit should be operated through a normal cycle to make sure it is functioning properly.

22.10 TERMS TO KNOW

anticipator
gas valve
oil burner
barometric damper

23.1 INTRODUCTION

So far this book has covered conventional types of heating systems. Heat pumps are discussed here because they are commonly used as heating units, and because their high operating efficiency will certainly cause an increase in that use.

A heat pump is a combination cooling-heating unit. Heat pumps generally are used in applications where the primary need is for cooling, but some heating also is required. A heat pump is especially efficient as a heating unit. As known sources of fuels and energy used for heating are becoming scarce, the use of heat pumps as both a primary and secondary source of heat has become important.

In this chapter, sections, components, and parts of a heat pump are described, and the operation of heat pumps is explained. The various types of heat pumps available are covered, as well as their application.

23.2 HEAT SOURCE

A heat pump, unlike a typical heating unit, does not produce heat. It simply transfers heat from one place to another. For a heat pump to be practical, heat must be available for transfer. Any material with a temperature above absolute zero has heat in it, however, so sources are common. Soil, **groundwater,** and the air in our atmosphere are all good potential sources of heat. Figure 23-1 shows a typical application using wells as a source of water for a heat pump system, and as a place to get rid of the water after use.

A typical heat pump used for heating buildings operates on the same principle as does an air conditioning system used to cool a building. In an air conditioning unit heat is picked up inside a building and expelled outside. A heat pump picks up heat outside and expels it inside. A heat pump extracts heat from a source, called a **heat sink,** and expels the heat where it can be used for heating a building.

The two most common sources of heat for a heat pump system are outside air and groundwater. In the winter, the outside air temperature ranges from below zero in some localities to as high as the inside design temperature. In almost all habitable areas, the average air temperature is

CHAPTER

23
Heat Pumps

Groundwater – Water from an underground source, such as a spring or a well.

Heat sink – Source of heat for a heat pump system, usually outside air or groundwater.

Well-water heat-sink
and waste

Indoor
section of
heat pump

Outdoor section
of heat pump

Injection
well

Supply
well

FIGURE 23–1
Many water source heat pumps use well water as a heat sink, and also return waste water to the soil through a well

higher than 10°F at any time of the year, while groundwater tends to remain about the same temperature all year, usually around 53°F. A typical heat pump will work well with a heat sink temperature down to about 0°F or lower. Groundwater always is a good source of heat for a heat pump.

23.3 SAVING ENERGY WITH A HEAT PUMP

Because picking up heat from a heat sink and moving the heat to a location where it is wanted is less expensive than generating that heat by a combustion process or with electricity, a heat pump is a very efficient device for heating. It saves both energy and money.

Since the heat in outside air or groundwater is available without direct cost, transferring this heat is less expensive than generating heat by the combustion process

FIGURE 23–2 ─────────────────────────────────
Use of a heat pump as a heating unit can mean savings as high or higher than three to one over some other types of energy

or by electric heating elements. The only cost is the cost for the energy to operate the heat pump. Electricity converts directly to 3.41 Btu of heat per watt of electricity used in a resistance heating element. Much more heat than this can be transferred by the use of 1 W when it is used to operate a heat pump. This savings in energy to produce heat can be as high as 3.5 to 1. In other words, 1 W of energy will produce the same amount of heat as 3.5 W would in a conventional system. Most of this heat is extracted from the heat sink. Figure 23-2 shows in graph form the savings in energy realized by using a typical heat pump system.

23.4 OPERATION

An understanding of how a heat pump works requires an understanding of the refrigeration cycle. In a refrigeration system, a chemical compound called a **refrigerant** is used to transfer heat. A refrigerant is a liquid with a boiling point temperature below that of normal **ambient** temperatures at atmospheric pressure. The refrigerant boils, or turns into a gas, at normal room temperature if not kept under pressure.

Refrigerant*

Ambient – "Surrounding" such as the temperature in the surrounding air.

*Term defined in text.

FIGURE 23–3
Boiling refrigerant. When a refrigerant boils, or changes state, heat is absorbed.

FIGURE 23–4
Condensed refrigerant. When a refrigerant condenses, heat is given up.

Expansion device – Device used in a refrigeration system to meter the flow of refrigerant in such a way that a high pressure is maintained on one side of the device and a low pressure on the other.

Evaporator – Coil in an air-conditioner or heat pump in which the refrigerant evaporates to produce cooling.

Condenser*

Air-to-air system*

Reversing valve*

The boiling point temperature of any liquid is directly related to the pressure exerted on it. If the pressure on a liquid is increased, the boiling point temperature of that liquid goes up; if the pressure is reduced, the boiling point temperature goes down. By controlling the pressure on the refrigerant in a refrigeration system, the boiling point temperature is controlled.

If the boiling point temperature of a refrigerant in a container is lowered enough by reducing the pressure, the boiling point temperature of the refrigerant can be made lower than the air around the container. Because of the temperature difference, heat will flow from the air around the container into the refrigerant, as shown in Figure 23-3. This causes the refrigerant to boil, or change to a gas, and as a result the surrounding air is cooled. Conversely, if the pressure on the refrigerant in a container is increased, the boiling point temperature of the refrigerant is raised higher than the surrounding air. Heat flows into the air from the refrigerant. See Figure 23-4. At this point, the refrigerant is in the vapor state. As the refrigerant gives up heat, it condenses into a liquid. Because the refrigerant gives up heat to the medium surrounding it, the medium is heated.

In a typical refrigeration system, pressure is reduced on the refrigerant in that part of the system where it is desired to pick up heat. The boiling point temperature is lowered below that of the air or water, and heat is picked up by the refrigerant. The pressure on the refrigerant is raised in that part of the system where the heat is to be rejected. In this case, the temperature of the refrigerant is raised above that of the air or water in that part of the system, and heat flows from the refrigerant to that air or water. This process is the typical refrigeration cycle.

An operating refrigeration system has both a high pressure side and a low pressure side. Coils are used for heat exchangers. One coil has refrigerant at a high pressure flowing through it, and another has refrigerant at a low pressure. See Figure 23-5.

Pressure is maintained on the high side of the system by a compressor. It is controlled on the low side by an **expansion device.** The coil on the low pressure side of the system is called an **evaporator** and is used to cool the air or water that is used to cool a building. The coil on the high side of the system is called the **condenser** and has air or water flowing over it to carry away the heat. Figure 23-6 shows a diagram of a typical air-to-air system. An

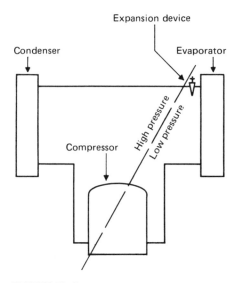

FIGURE 23–5 ⎯⎯⎯⎯⎯⎯⎯⎯⎯⎯⎯⎯⎯
Diagram of a refrigeration system showing the high pressure side and the low pressure side

FIGURE 23–6 ⎯⎯⎯⎯⎯⎯⎯⎯⎯⎯⎯⎯⎯
Diagram of a typical air-to-air conditioning system

air-to-air system is one in which the air is used as both a heat source and as a medium to carry the cool air.

When the system is used as a heat pump, the relative position of the two coils is reversed by changing the pressures in the two sides of the system. In the heating mode the coil that is rejecting heat in the air-conditioning cycle is now collecting it, and the other coil is giving up the heat. On an air-to-air system, heat is picked up by the outside coil from the outside air, and is rejected inside a building by the inside coil. The heat is then used to heat the building. A refrigerant control device called a **reversing valve** is used to reverse the direction of refrigerant flow in a heat pump system. Figure 23-7 shows the relative position of the indoor and outdoor coils and the location of the reversing valve.

Heat pumps usually are built so they can be used for heating or cooling. They have controls and devices that make it possible to reverse the direction of the refrigerant flow, which changes the pressure in the two sides of the system. The heat pump can operate as an air-conditioner in one mode of operation and as a heater in the other. A heat pump can be called a reverse-cycle air-conditioner.

FIGURE 23–7 ⎯⎯⎯⎯⎯⎯⎯⎯⎯⎯⎯⎯⎯
Diagram of heat pump system showing reversing valve

FIGURE 23–8 ─────────────
Major components of an air-conditioning system

FIGURE 23–9 ─────────────
Heat pump system showing the major parts

23.5 OPERATING DEVICES AND CONTROLS

A refrigeration system, or air-conditioning system, has six major components. They are

1. Evaporator coil
2. Condenser coil
3. Compressor
4. Expansion device
5. Refrigerant lines
6. Controls

These major components are shown on the diagram in Figure 23-8.

A heat pump has these same major components, but some additional parts are used to make it operable in the reverse cycle. These additional parts are

1. Reversing valve
2. Additional expansion device and check valves, or other expansion devices
3. Heat-cool thermostat
4. Defrost controls

These parts are shown on the diagram in Figure 23-9. Each will be discussed in the section that follows.

Operating Devices

The reversing valve and the additional expansion device with the **check valves** are needed to make a heat pump operate in the heating mode. The reversing valve is the device that actually reverses the direction of flow the refrigerant in the system. The added expansion device and the check valves are used to change the point in the system where the pressure changes from high to low.

Reversing Valve. To change a heat pump operation from cooling to heating, it is necessary to reverse the direction of flow of the refrigerant. The compressors used in refrigeration systems, however, are nonreversible; that is, the refrigerant goes through them in only one direction. It always goes in one port, called the **suction port,** and comes out the other, called the **discharge port.** Another device is needed to reverse the flow of the refrigerant. This part is the reversing valve. The reversing valve receives the refrigerant from the discharge port of the compressor, and routes it to one or the other of the coils, depending upon whether cooling or heating is wanted. Figure 23-10 shows a reversing valve. The valve also takes the return vapor from the opposite coil and routes it to the compressor suction port.

Most reversing valves are piston type. A piston is moved back and forth inside a cylinder to control the direction of refrigerant flow. During the operation of the reversing valve movement of the piston opens or closes ports to the refrigerant lines in the system according to its position in the cylinder. The piston and the line ports can be seen in the photograph in Figure 23-11.

The reversing valve is a **pilot-operated valve.** An electric solenoid valve in the pilot valve controls pressure on the ends of the cyinder in the main valve. The main valve piston is moved by the pressure, which is exerted by the refrigerant from the operating system. Figure 23-11 also shows the pilot valve.

Check valve – Mechanical device placed in a pipe that will allow liquids or gases to flow in only one direction in the pipe .

Suction port*

Discharge port*

Pilot-operated valve – Valve operated by system pressure that is controlled by a smaller valve.

FIGURE 23–10 —————————
Typical reversing valve used in a residential heat pump

FIGURE 23–11 —————————
Cutaway photograph of a reversing valve showing the operating parts and the pilot-operated valve

Compressor discharge

Valve body

Piston

Outdoor coil

Indoor coil

Compressor suction

FIGURE 23–12

Diagram of a reversing valve showing the slide in the cooling position

Compressor discharge

Piston

Valve body

Outdoor coil

Indoor coil

Compressor suction

FIGURE 23–13

Diagram of a reversing valve showing the slide in the heating position

The solenoid of the pilot valve is energized by the heat pump control system, usually a thermostat. On a call for cooling, a small piston in the pilot valve opens a port to a capillary tube that runs from the pilot valve to one end of the main reversing valve. This capillary tube connects the one end of the main valve cylinder to the suction side of the refrigeration system through the pilot valve. Pressure differential between the two ends of the cylinder moves the piston in the main valve toward this end of the cylinder. The ports in the main valve now line up so the suction line from the compressor is connected to the refrigerant line from the inside coil, and the discharge line from the compressor is connected to the refrigerant line to the outside coil. In this configuration, the unit is in the cooling mode. Figure 23-12 shows a reversing valve in this piston.

On a call for heat, the solenoid valve in the pilot valve is energized. This moves the piston in the pilot valve to the opposite end of that described for cooling. When the pilot valve piston is in this position, a capillary tube running from this end of the pilot valve cylinder to the end of the master reversing valve cylinder from that described for cooling is connected to the low pressure side of the refrigeration system through the pilot valve. This causes a reduction in pressure in this end of the main valve cylinder. The reduction in pressure causes the main valve piston to move toward this end of the reversing valve. When the piston is in this position ports in the valve are lined up so the suction line from the compressor is connected through the main valve to the refrigerant line from the outside coil, and the discharge line from the compressor is connected to the refrigerant line going to the inside coil. This configures the unit in the heating mode. Figure 23-13 shows a reversing valve in this position.

Expansion Devices. An expansion device is a device used in a refrigeration line in an air conditioning system to produce a given pressure drop in the system, and to meter the flow of refrigerant. A compressor is used to raise the pressure of the refrigerant vapor in a system so heat can be removed in the evaporator coil. An expansion device is used to maintain that pressure in the high pressure side of the system, and reduce the pressure to that needed in the low pressure side of the system.

Expansion Valves. If an air-conditioner uses an expansion valve as an expansion device, the valve is located on the liquid line where the line enters the evaporator coil. On a heat pump, the function of the indoor and outdoor coils is reversed when the system is operating in the heating mode. An expansion device used for the cooling cycle will not work on the heating cycle. To make the system workable for both cooling and heating, either an additional expansion device is needed on the outdoor coil, or some type of expansion device is needed that will work when the refrigerant flows in either direction. One answer is the use of an expansion valve on each coil. A check valve is installed in bypass piping around each valve, to force the refrigerant through the valve for each mode of operation. Other types of refrigerant control devices are capillary tubes, pressure drop refrigerant lines, and modulating flow control check valves.

Check Valves. A refrigerant will flow through a conventional expansion valve in only one direction. In a heat pump using expansion valves for expansion devices, check valves are used in bypass circuits that put the check valves in parallel with the expansion valves. This allows refrigerant to flow correctly through each of the coils, for each mode of operation. The check valve in the bypass circuit at the indoor coil causes the refrigerant to pass through the valve on a cooling cycle, but allows it to bypass the valve on a heating cycle. The check valve on the outdoor coil causes the refrigerant to flow through the valve on a heating cycle, but allows it to bypass the valve on a cooling cycle.

See Figure 23-14 for the location of the bypass and the extra expansion device.

Other Expansion Devices. If an expansion device is used that allows refrigerant flow in either direction, the device must provide a restriction in either direction so the refrigerant going through it will expand. It must also provide for more restriction in one direction than the other because of the difference in refrigerant pressure in the heating mode from the cooling mode.

One type of refrigerant flow control used with smaller heat pumps is a **capillary tube.** The pressure drop through a capillary tube expansion device is basically the same in either direction. A capillary tube used as an expansion device on a small heat pump system is shown in Figure 23-15.

FIGURE 23–14 _____
Diagram showing the heat pump system with the extra expansion device, the refrigerant by passes, and the check valves

Air — Outdoor unit

Capillary tube

Reversing valve

Air — Indoor unit

Heat pump
Air to air
cap tube

Compressor

FIGURE 23–15 _____

Many small heat pump systems with the outdoor and indoor units close to each other use a capillary tube expansion device

A **precisely sized liquid line** between the indoor and outdoor coils that provides enough pressure drop so it functions as an expansion device regardless of which way the refrigerant flows thorugh it, is a type of refrigerant flow control used on some medium-sized heat pumps. The liquid line is sized in both diameter and length to provide the proper amount of pressure drop for refrigerant expansion.

Another expansion device used on heat pumps is a **metering valve** and **check valve combination**. This device has a precisely sized orifice through which the refrigerant flows in one direction. When refrigerant flow is reversed, a ball check in the valve closes off most of the opening in the other direction, but leaves enough of an opening so some refrigerant can still get through. The opening size for flow in this direction is again precisely sized for the proper pressure drop. An expansion device of this type is shown in Figure 23-16.

Controls

In addition to the normal controls used on air-conditioning units, a heat pump must have a heat-cool thermostat and

FIGURE 23–16
Various types of flow control devices are used as reversing valves for heat pumps

controls to defrost the outside coil when frost or ice builds up on it.

Heat-Cool Thermostat.   Because a heat pump is a unit that can provide both heating and cooling, a thermostat to control its operation must be used that will call for either heat or cooling as needed. This can be accomplished in one of several ways. Most commonly, either three- or four-bulb thermostats are used. A three-bulb thermostat has three mercury bulb switches in it, while a four-bulb thermostat has four mercury bulb switches. Usually, when a three bulb thermostat is used one bulb controls the compressor for both heating and cooling, one bulb controls the reversing valve, and the other bulb controls auxiliary heat.

When a four-bulb thermostat is used, each bulb in the thermostat controls one part of the heat pump operation. Two bulbs control the cooling and two bulbs the heating. A heat pump thermostat is shown in Figure 23-17.

FIGURE 23–17
Thermostat used on a heat pump system

With a four-bulb thermostat, the system normally is in the heating mode. When the first stage of cooling is called for, the reversing valve switches to the cooling position. The second-stage cooling bulb then brings on the compressor. When the thermostat is in the normal heating position, the first-stage heat bulb energizes the compressor, and if the second-stage heat bulb calls for more heat, the thermostat brings on the auxiliary heat. This sequence is shown in the block diagram in Figure 23-18.

Defrost Controls. When a heat pump operates in a heating cycle, the outside coil is acting as an evaporator. The temperature of the refrigerant in the outdoor coil often will be below 32°F when the unit is operating in the heating mode. When the temperature falls below 32°F, frost or ice can form on the coil from the moisture that condenses on it from the air. If this frost or ice starts to build up, the amount of air that can be blown across the coil is reduced, and the refrigerant temperature drops still more. Eventually, the unit will go off on low pressure, or damage will occur to the compressor.

To prevent this buildup of ice, controls must be added to the unit to monitor the icing condition, and some defrost system must be used to melt the frost or ice. Various methods may be used for defrost. Electric heating elements, warm air or even water may be used for some types of defrost systems. Most commonly, however, hot

FIGURE 23–18
Block diagram showing the heat pump control components in relation to the thermostat

refrigerant gas from the refrigeration system is used. Common to all defrost systems are the controls to sense the degree of frosting or icing, starting the defrost cycle, and the timing when the defrost cycle should be terminated. These are called **defrost controls.**

In a system using hot gas defrost, hot refrigerant gas is cycled through the coil so defrosting takes place. To get hot gas to flow through the outdoor coil, the heat pump system operated in the cooling mode. To prevent cold air from being blown into the building during the defrost cycle, the defrost controls also bring on the **auxiliary heat.** The defrost cycle lasts only long enough to accomplish the melting of the frost or ice. As soon as the controls sense that the coil is free of ice or frost, they put the unit back into the heating mode. This sequence is shown on the block diagram in Figure 23-19.

23.6 TYPES OF UNITS

Two major types of heat pump systems are used for heating buildings: (1) the **compatible-component type unit,** in which the components of the system are furnished by a manufacturer so they work together, and (2) the **built-up type unit,** in which the components are selected and the system designed by an engineer. The first type almost always is used on residential buildings and smaller commercial buildings, and the capacities are seldom very large. The second type is most often used on larger buildings, and the capacity of the system usually is much greater. Within these two major types, other divisions are found concerning operating characteristics, physical characteristics, and heating-cooling capacity of the units.

There are four different possibilities for operating characteristics. These characteristics can be listed by the heat source and the heating-cooling medium

1. A unit that uses water as a heat source and water as the heating-cooling medium
2. A unit that uses water as a heat source but air as the heating-cooling medium
3. A unit that uses air as a heat source and water as the heating-cooling medium
4. A unit that uses air as a heat source and also as the heating-cooling medium

FIGURE 23-19
Block diagram showing the components of the defrost system in relation to each other

Defrost controls*

Auxiliary heat – Heat that is added in a heat pump system by means other than the heat pump.

Compatible-component type unit*

Built-up type unit*

Package unit*

Split system*

Ton of cooling*

The division by physical characteristics includes the **package units,** which have all of the components in one sheet metal box, and the **split systems,** which have the components located in different parts of the building or even outside the building.

The division by equipment capacity will be discussed in sections by categories.

Residential and Small Commercial Units

Heat pumps for small commercial and residential buildings are nearly all the compatible-component type unit— that is, they have components that are meant to work together. Some are split systems, and some are packaged units. Figure 23-20 shows an outdoor unit used with a split system heat pump. The indoor unit is a coil mounted on the furnace or a blower coil unit in the building. The two parts of the system are connected by refrigerant lines.

The heating capacity of package heat pump units and compatible-component units is somewhat limited. The

FIGURE 23–20
Outdoor unit of a split-system heat pump

units are rated from 1 ton to about 15 to 20 tons capacity. This measurement is cooling capacity. One **ton of cooling** is equivalent to 12,000 Btu/h. The heat capacity of a heat pump is more difficult to state because the output varies as the temperature of the heat source medium varies. Generally, the heating capacity of an air source heat pump, when the outside air temperature is 45°F, is about the same as the cooling capacity of the unit at design conditions. Heat pumps are rated at 45°F for high end and 15-17°F for low-end capacities.

Package Heat Pumps.   Package heat pumps are completely self-contained units. All of the components are contained in one sheet metal cabinet. Each heat pump is built so it only needs to be connected to a source of water, with a drain, if it is a water source unit, and to the ductwork for the distribution system. Or, if it is an air source unit, it only needs to be installed outside the building with connections to the ductwork. Figure 23-21 illustrates a package unit.

Some smaller package units are built to be installed in windows, with part of the unit outside and part inside, or through an outside wall with the same arrangement. These units are quite small. Their maximum capacity is about

FIGURE 23–21
A package heat pump has both outdoor and indoor coils in one package

FIGURE 23–22
Window-model heat pump

two tons of air-conditioning and proportionate heating ability. These units are meant to be used for conditioning one room at a time. Figure 23-22 illustrates a typical window unit.

Larger package heat pumps-from about two tons to 15 tons of capacity are designed to be used with duct systems.

Split Systems. Many split-system heat pump units use outside air as a source of heat. The outside part of the system includes the outside coil, a blower to move the air across the coil, and a compressor, with the controls for these components. The inside part of the system includes the inside coil and a blower to move the conditioned air. Split systems are available with indoor coil units to be used with remote blowers and also with blower coil units. Split systems usually are used in applications where it is not convenient to duct air into a building from the outside.

Units for Larger Buildings

Heat pump systems that need more heating for larger building are often the built-up type of unit, that is, the components are selected separately. Water usually is the source of heat and also the medium used to carry the heat to the point of use. Large multistory buildings and buildings heated and cooled from a central station system are included in this category. The capacities of these units may be in the hundreds of tons.

23.7 SUMMARY

A heat pump is a refrigeration unit that collects heat and moves it from a place where it is not needed to a place where it is needed. A heat pump works on the same principle as any refrigeration unit. When the pump is operating in the heating mode, it can be called a reverse-cycle air-conditioning unit.

A heat pump is a more sophisticated piece of equipment than a gas, oil, or electric furnace. It also makes more heat available per unit of energy used than any of the furnaces. Because of its high efficiency, it is fast becoming one of the most popular units used for heating buildings.

Manufacturers of heat pumps build units in a wide

variety of types and capacities. Self-contained units and compatible component units, to be used as split systems, are available from one ton of air-conditioning capacity up to 15- or 20-ton capacity. The components to build up larger systems are available for almost any capacity desired.

23.8 TERMS TO KNOW

groundwater
heat sink
refrigerant
ambient
expansion device
evaporator
condenser
air-to-air system
reversing valve
check valve
suction port
discharge port
pilot-operated valve
capillary tube
precisely sized liquid line
metering valve
defrost controls
auxiliary heat
compatible-component type unit
built-up unit
package unit
split system
ton of cooling

CHAPTER

24

Heat Pump Sizing and Application

24.1 INTRODUCTION

Sizing a heat pump as a heating unit, as opposed to sizing it for a cooling unit, requires consideration of the fact that a heat pump using outside air as a heat sink has a heat output that goes down as the outside temperature goes down. This is just the opposite effect of that desired, since heat loss of a building goes up as the outside temperature goes down. In a warm climate where the demand for cooling is quite high, heat pumps usually are sized for the cooling load. In climates where the heating demand is quite high in the winter, heat pumps are sized to provide the heat required to heat a building down to a given outside temperature, called the balance point temperature, and auxiliary heat is used to supplement the heat pump output below that temperature.

24.2 HEAT OUTPUT vs. TEMPERATURE

Both the heating output of a heat pump, and the heat loss of a building, vary with outside temperature. With the two variables, heat output and heat loss, being plotted against the one variable, outside temperature, a correlation can be drawn between output and loss. This correlation can be expressed either in two graphs or in one graph. If two graphs are used, one graph shows the heating output of one or more heat pumps plotted against outside temperature. The second graph shows the heat loss of a building plotted against outside temperature. If one graph is used both graph curves are plotted against outside temperature. The heat output curve shows heat output going down as the temperature goes down, and shows the heat loss going up as the temperature goes down.

The graph in Figure 24-1 shows the heating capacity of various heat pumps at varying outside temperatures. The Btu/h output of the pumps is shown in Btu/h on the scale on the left of the graph. Outside design temperatures are shown on the scale across the bottom. The heat output of the pumps is shown on the diagonal lines coming down from the right upper corner. Where any heat pump line crosses a vertical temperature line, the Btu/h output of the unit at that temperature is read on the Btu/h scale at the left. Each of the heat pump output curves is for a different capacity unit.

FIGURE 24-1
Typical capacity curves for four sizes of heat pumps

24.3 SIZING

To size a heat pump for heating a building in a climate in which the heating load on a building is quite high and the cooling load quite low, the heat output of the heat pump unit should be compared to the heat loss of the building it will be used in. To do this, plot both building heat loss and heat pump output on one graph. A graph should be prepared showing outdoor temperatures on the X axis across the bottom of a graph, and with heat loss in Btu shown on the Y axis up the left side of the graph. The heat loss for a specific building is shown on the graph by drawing a line from the point where the indoor design temperature used for the heat loss calculations intersects with 0 Btu/h, up to the point where the line running across from the heat loss for the building at the outdoor design temperature intersects with the outdoor design temperature used. This can best be shown in Figure 24-2.

By adding the heat loss curve to the heat output curves for various heat pumps, a designer can identify the **balance point** for a job.

The point where the heat pump output curve crosses the heat loss curve is called the balance point of the system. The curves for a typical job are shown on Figure 24-3. The heat pump for a particular job can be selected by picking the unit that will give the most desirable balance point.

Balance point*

*Term defined in text.

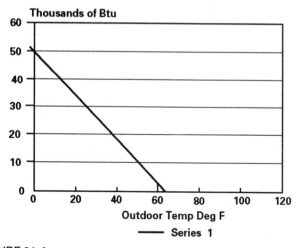

Heat loss - Btu / Deg F

FIGURE 24–2
Heat loss for any building may be graphed by plotting the loss in Btu/h against outdoor temperatures

Balance Point. The balance point for any job is related to the heat loss curve for that job only, and the heat output curve for a specific heat pump. Since the heat output of a heat pump goes down as the outside temperature goes down, the efficiency of the unit also goes down with the temperature. It is important to select a unit that will operate as efficiently as possible, yet is not too large. The cost of a heat pump is relative to its size and capacity. A balance point of about 30°F to 35°F is right for most jobs. When the heat loss curve is drawn on the graph, a unit should be selected that gives a balance point temperature in that range.

EXAMPLE 24-1

A building to be constructed in Seattle, Washington, where the outside design temperature is 15°F, has a heat loss of 38,600 Btu/h at design conditions. Select a heat pump for that building.

Solution

Draw the heat loss curve on the balance point chart, as shown in Figure 24-4. Look for the heat pump output curve that crosses the heat loss line closest to 35°F outside

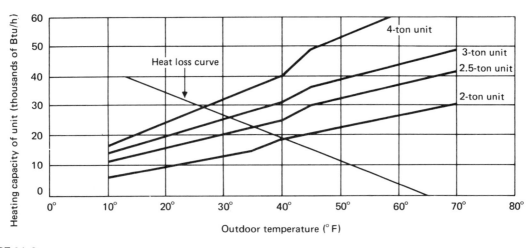

FIGURE 24–3
Heat loss for a given building plotted against the capacity curves for a selection of heat pumps

temperature. For this job, a 2.5-ton heat pump should be used.

24.4 COEFFICIENT OF PERFORMANCE (COP)

Since the heating output of a heat pump varies with the temperature of the heat source, but the amount of energy

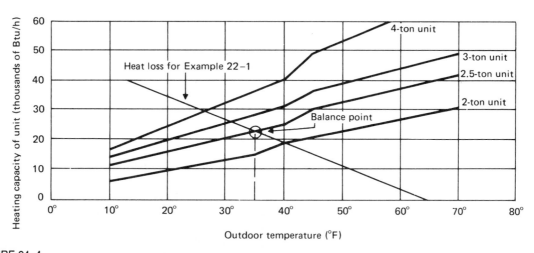

FIGURE 24–4
Balance point chart for the building described in the text

required to operate the unit does not vary significantly, the heating efficiency of the unit varies with the temperature also. The efficiency of the unit is found by dividing the heat output by the heat equivalent of the energy used to operate the unit. The resulting figure is called the coefficient of performance (COP) of the unit:

$$COP = \frac{Btu/h \times output\ of\ unit}{Btu/h \times input\ of\ energy}$$

This COP figure will vary from about 1.5 at the lowest outside operating temperatures to a high of about 3.5 at the highest temperatures.

For an electric heat pump, the heat equivalent of the energy used to operate the unit is provided in the manufacturer's specification sheets. The wattage to operate the unit at various outside temperatures is shown there with the heating capacity for the same conditions. The wattage is multiplied by the factor of 3.41 Btu/W to find the Btu in heat for the electric energy.

EXAMPLE 24-2

A heat pump produces 34,500 Btu/h of heating at a given set of conditions and requires 3679 W to operate at that set of conditions. Find the COP.

Solution

$$COP = \frac{Btu/h\ produced}{Btu/h\ used}$$

$$= \frac{35,400}{3679 \times 3.41}$$

$$= 2.75$$

The COP of the unit at the operating conditions is 2.75.

The COP of a family of typical heat pumps at various outside temperatures is shown in Figure 24-5.

Seasonal Heating Performance Factor (SHPF).   Another term, seasonal performance factor, often is used to provide a guide to the performance of a heat pump with factors

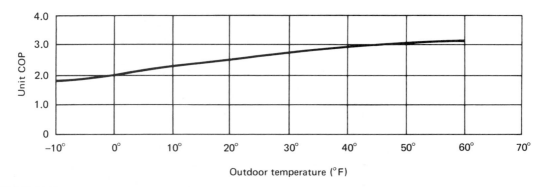

FIGURE 24–5
The COP of heat pumps varies as the outdoor temperature varies

built-in regarding seasonal performance. Some of these factors relate to performance variations due to starting and stopping, and are adjusted for defrost activities. Manufacturers provide SHPF numbers for their units operating at various temperatures and conditions.

24.5 AUXILIARY HEAT

The heating capacity of a heat pump is enough to heat the building only down to the balance point temperature. Any time the outside temperature goes lower than the balance point temperature, extra heat is required to heat the building. This extra heat, called auxiliary heat, may be supplied by any other source. In most cases, electric resistance units are used as an auxiliary heat source.

Most heat pumps automatically cut-out somewhere around 0°F, or slightly less. The low pressure cut-out on the unit will shut off the operation if the outside temperature gets below that point. Some heat pumps are furnished with an outdoor thermostat that will keep the compressor from running if the outdoor temperature gets below its shutoff point. Thus, if the design temperature used for the heat loss is higher than 0°F, theoretically only enough heat needs to be supplied by the auxiliary heat to make up the difference between the output of the pump and the heat loss of the building at the design temperature. See Figure 24-6.

If the design temperature used in the heat loss figures is less than 10°F, enough auxiliary heat should be furnished to provide the total amount of heat for the building. See Figure 24-7.

FIGURE 24-6

The amount of auxiliary heat needed for heating a building can be plotted on the balance point graph

In practice, it is better to provide enough auxiliary heat to carry the entire heating load at design temperature in any case. Enough heat must be available during defrost cycles and in case the heat pump is shut down for service.

24.6 APPLICATION

The design of a heat pump system requires more attention to certain application details than does the design of a conventional heating system. Because a heat pump is a refrigeration unit, the quantity of air or water used as the heat source is critical. If the cfm of air or the gpm of water on either side of the system drops too low, the temperature of the refrigerant in the system will increase, and the unit will either cycle off on high pressure, or damage to the compressor will be sustained. The designer must make sure that there is an adequate supply of air or water to each side of the system and that the ducts or piping, blowers, and pumps are sized correctly.

On the heat sink side of the system, an adequate source of air or water must be available at a high enough temperature to allow the extraction of heat at a constant rate. The source must be such that the air or water can be used with

FIGURE 24–7

Amount of heat required as auxiliary heat if the design temperature of the job is less than the cut-out temperature for the heat pump

the equipment that the designer chooses for the job.

On the heat distribution side of the system, the ducts or piping and the blowers or pumps must be designed so the quantity of air or water to carry the heat through the building will not vary.

The distribution system that carries the air or water must be designed so the building will be heated properly.

Heat Source. The heat sink, or source of heat for the heat pump, can be air or water. If it is air, the usual practice is to use the outside air. If it is water, the usual sources are public system, stream or lake, or wells.

A heat pump system that uses air as its heat source is relatively simple to design. The part of the system containing the outside coil needs to be located outside the building so it has a source of clean air. This part of the system contains a blower to draw the air across the coil. In some cases, a designer may be tempted to install the outside part of a unit inside the building and duct the air to it from the outside. Unless the unit is especially designed for this application, this placement should not be done. The air mover in a typical unit does not have a blower that will move air against the resistance of the ductwork, and the unit will be under aired if it is installed in that way.

Two precautions should be taken in locating the outside unit: (1) the discharge air from one unit should not enter the intake of another, and (2) the discharge air from any unit should not recycle back into its own entering air side. In either case, the temperature of the entering air would be too low, and the heating capacity of the unit would be reduced.

Water source heat pumps usually are used in one of two situations: when there is a plentiful, cheap supply of water available or when the unit or units need to be installed inside a building where it is difficult to get air to them from outside. Water is a good choice in both cases because it is a good source of heat for a heat pump. The temperature of the water is normally quite stable, and the output of the unit does not vary greatly.

When public utility water is used as the heat source, the cost of the water can be an important factor. In some cases, the waste of the water is used as the heat source.

When public utility water is used as the heat source, the cost of the water can be an important factor. In some cases, the waste of the water after it is used is more important. In most large U.S. cities, the waste of large quantities of water is not acceptable.

Lake or stream water that is used must be from a location where the quantity does not vary. In most cases, the water from either source should be filtered and possibly may have to be treated for contaminants before it is used. The water will normally have to be pumped from the source to the point of use.

If the size of the installation warrants the cost, the use of well water may be the best answer to the source of water for a heat pump system. When wells are used, usually at least two are drilled. One well is used for the source of the water, and the other is for the waste of the used water. The wells are spaced far enough apart so the waste water will mix with groundwater before entering the supply well again. Usually the system is designed so the use of the two wells can be alternated to control the water level that supplies the wells.

In commercial applications, water source heat pumps often are used in conjunction with a run-around water piping system. This system often is used in buildings with some zones that need cooling while other zones need heating. Water is piped from one zone to another, and in the zone needing cooling heat is removed from the water to provide the cooling, and then piped to the part of the building needing heating. There, the water is used for heating. This type of system is often used with a variable-volume (VAV) type of heating/cooling system.

Heat Distribution. On the distribution side of the system, either air or water can be used as the medium to carry the heat. When air is used, it is normally the same air as is circulated through the building for heating. When water is used, it is distributed to terminal units where a heat exchanger is used to heat air, which is then used to heat the building.

When air is the medium, the temperature rise through the inside coil is much lower than that for any other type of heating system, probably in the range of 15°F to 20°F. The air introduced into the building through the registers will be much cooler than that from a fuel-fired or electric heating unit. Therefore, the registers must be located very carefully to prevent the relatively cool air from blowing on occupants of the building. Air at a discharge temperature normal with heat pumps blowing directly on a person will feel like a cold draft.

When a heat pump is used on a hydronic heating system, the main concern is that the entire system be properly sized for the quantity of water required through the heat pump, and for the temperature of water available from the heat pump.

24.7 SUMMARY

Special consideration has to be given to the application of heat pumps. In warmer climates, where air conditioning is the main concern, heat pumps are used primarily as air conditioning units with the heat output being adequate for the heating load on a building. In colder climates, where heating is the main concern, the heat output of a heat pump must be determined accurately for all outside conditions, since the unit can be used as a primary source of heating, with the cooling output as a bonus during warm weather.

All heat pump systems need an adequate heat sink, and the distribution system must be designed with a constant supply of air or water for distribution of the heat supplied.

24.8 TERMS TO KNOW

balance point
coefficient of performance (COP)
seasonal heating performance factor (SHPF)
auxiliary heat
heat source
heat distribution

Appendix Tables

TABLE A–1 _____
Weight and Heat Value for Some Fuel Woods

Species	Btu/lb[a]	Weight/cord[b]	Btu/cord[c]
Alder	6720	2540	17,068,800
Apple	7890	4400	34,716,000
Ash	8300	3440	28,552,000
Aspen	5980	2160	12,916,800
Birch, white	7470	3040	22,708,800
Cedar, western	6720	2060	13,843,200
Cherry	7470	3200	23,904,000
Cottonwood	5980	2160	12,916,800
Dogwood	8300	4230	35,109,000
Elm	7470	2260	16,882,200
Fir, Douglas	7470	2970	22,185,900
Fir, grand	5980	2160	12,916,800
Hemlock	6720	2700	18,144,000
Hickory	9500	4240	40,280,000
Juniper	7470	3150	23,530,500
Larch, western	7890	3330	26,273,700
Madrone	8300	4320	35,856,000
Maple, red	7885	3200	25,232,000
Maple, sugar	7890	3680	29,035,200
Oak, red	8300	3680	30,544,000
Oak, white	9500	4200	38,900,000
Pine, lodgepole	5980	2610	15,607,800
Pine, ponderosa	6720	2240	15,052,800
Pine, white	6720	2250	15,120,000
Pine, yellow	7890	2610	20,592,900
Poplar	5980	2080	12,438,400
Redwood	7470	2400	17,928,000
Spruce	5980	2070	15,462,900
Spruce, Norway	5980	2240	13,395,200

[a] Approximate heat value, Btu/lb, for air-seasoned (20% moisture) wood

[b] Approximate weight, lb/cord, for average density of air-seasoned wood

[c] Approximate Btu/cord for air-seasoned (20% moisture) wood and for average density

TABLE A-2

Orifice Sizing Chart for Utility Gas (Cubic Feet per Hour at Sea Level)

Specific Gravity = 0.60
Orifice Coefficient = 0.9
For utility gases of another specific gravity, select factor from Table 5,
For altitudes above 2,000 feet, first select the equivalent orifice size at sea level
    from Table 6.

Orifice Size (Decimal or DMS)	Gas Pressure at Orifice - Inches Water Column								
	3	3.5	4	5	6	7	8	9	10
.008	.17	.18	.19	.23	.24	.26	.28	.29	.30
.009	.21	.23	.25	.28	.30	.33	.35	.37	.39
.010	.27	.29	.30	.35	.37	.41	.43	.46	.48
.011	.33	.35	.37	.42	.45	.48	.52	.55	.59
.012	.38	.41	.44	.50	.54	.57	.62	.65	.70
80	.48	.52	.55	.63	.69	.73	.79	.83	.88
79	.55	.59	.64	.72	.80	.84	.90	.97	1.01
78	.70	.76	.78	.88	.97	1.04	1.10	1.17	1.24
77	.88	.95	.99	1.11	1.23	1.31	1.38	1.47	1.55
76	1.05	1.13	1.21	1.37	1.52	1.61	1.72	1.83	1.92
75	1.16	1.25	1.34	1.52	1.64	1.79	1.91	2.04	2.14
74	1.33	1.44	1.55	1.74	1.91	2.05	2.18	2.32	2.44
73	1.51	1.63	1.76	1.99	2.17	2.32	2.48	2.64	2.78
72	1.64	1.77	1.90	2.15	2.40	2.52	2.69	2.86	3.00
71	1.82	1.97	2.06	2.33	2.54	2.73	2.91	3.11	3.26
70	2.06	2.22	2.39	2.70	2.97	3.16	3.38	3.59	3.78
69	2.25	2.43	2.61	2.96	3.23	3.47	3.68	3.94	4.14
68	2.52	2.72	2.93	3.26	3.58	3.88	4.14	4.41	4.64
67	2.69	2.91	3.12	3.52	3.87	4.13	4.41	4.69	4.94
66	2.86	3.09	3.32	3.75	4.11	4.39	4.68	4.98	5.24
65	3.14	3.39	3.72	4.28	4.62	4.84	5.16	5.50	5.78
64	3.41	3.68	4.14	4.48	4.91	5.23	5.59	5.95	6.26
63	3.63	3.92	4.19	4.75	5.19	5.55	5.92	6.30	6.63
62	3.78	4.08	4.39	4.96	5.42	5.81	6.20	6.59	6.94
61	4.02	4.34	4.66	5.27	5.77	6.15	6.57	7.00	7.37
60	4.21	4.55	4.89	5.52	5.95	6.47	6.91	7.35	7.74
59	4.41	4.76	5.11	5.78	6.35	6.78	7.25	7.71	8.11
58	4.66	5.03	5.39	6.10	6.68	7.13	7.62	8.11	8.53
57	4.84	5.23	5.63	6.36	6.96	7.44	7.94	8.46	8.90
56	5.68	6.13	6.58	7.35	8.03	8.73	9.32	9.92	10.44
55	7.11	7.68	8.22	9.30	10.18	10.85	11.59	12.34	12.98
54	7.95	8.59	9.23	10.45	11.39	12.25	13.08	13.93	14.65
53	9.30	10.04	10.80	12.20	13.32	14.29	15.27	16.25	17.09
52	10.61	11.46	12.31	13.86	15.26	16.34	17.44	18.57	19.53
51	11.82	12.77	13.69	15.47	16.97	18.16	19.40	20.64	21.71
50	12.89	13.92	14.94	16.86	18.48	19.77	21.12	22.48	23.65
49	14.07	15.20	16.28	18.37	20.20	21.60	23.06	24.56	25.83
48	15.15	16.36	17.62	19.88	21.81	23.31	24.90	26.51	27.89
47	16.22	17.52	18.80	21.27	23.21	24.93	26.62	28.34	29.81
46	17.19	18.57	19.98	22.57	24.72	26.43	28.23	30.05	31.61
45	17.73	19.15	20.52	23.10	25.36	27.18	29.03	30.90	32.51
44	19.45	21.01	22.57	25.57	27.93	29.87	31.89	33.96	35.72

TABLE A–2 *continued*

Orifice Size (Decimal or DMS)	3	3.5	4	5	6	7	8	9	10
43	20.73	22.39	24.18	27.29	29.87	32.02	34.19	36.41	38.30
42	23.10	24.95	26.50	29.50	32.50	35.24	37.63	40.07	42.14
41	24.06	25.98	28.15	31.69	34.81	37.17	39.70	42.27	44.46
40	25.03	27.03	29.23	33.09	36.20	38.79	41.42	44.10	46.38
39	26.11	28.20	30.20	34.05	37.38	39.97	42.68	45.44	47.80
38	27.08	29.25	31.38	35.46	38.89	41.58	44.40	47.27	49.73
37	28.36	30.63	32.99	37.07	40.83	43.62	46.59	49.60	52.17
36	29.76	32.14	34.59	39.11	42.76	45.77	48.88	52.04	54.74
35	32.36	34.95	36.86	41.68	45.66	48.78	52.10	55.46	58.34
34	32.45	35.05	37.50	42.44	46.52	49.75	53.12	56.55	59.49
33	33.41	36.08	38.79	43.83	48.03	51.46	54.96	58.62	61.55
32	35.46	38.30	40.94	46.52	50.82	54.26	57.95	61.70	64.89
31	37.82	40.85	43.83	49.64	54.36	58.01	61.96	65.97	69.39
30	43.40	46.87	50.39	57.05	62.09	66.72	71.22	75.86	79.80
29	48.45	52.33	56.19	63.61	69.62	74.45	79.52	84.66	89.04
28	51.78	55.92	59.50	67.00	73.50	79.50	84.92	90.39	95.09
27	54.47	58.83	63.17	71.55	78.32	83.59	89.27	95.04	99.97
26	56.73	61.27	65.86	74.57	81.65	87.24	93.17	99.19	104.57
25	58.87	63.58	68.22	77.14	84.67	90.36	96.50	102.74	108.07
24	60.81	65.67	70.58	79.83	87.56	93.47	99.83	106.28	111.79
23	62.10	67.07	72.20	81.65	89.39	94.55	100.98	107.49	113.07
22	64.89	70.08	75.21	85.10	93.25	99.60	106.39	113.24	119.12
21	66.51	71.83	77.14	87.35	95.63	102.29	109.24	116.29	122.33
20	68.22	73.68	79.08	89.49	97.99	104.75	111.87	119.10	125.28
19	72.20	77.98	83.69	94.76	103.89	110.67	118.55	125.82	132.36
18	75.53	81.57	87.56	97.50	108.52	116.03	123.92	131.93	138.78
17	78.54	84.82	91.10	103.14	112.81	120.33	128.52	136.82	143.91
16	82.19	88.77	95.40	107.98	118.18	126.78	135.39	144.15	151.63
15	85.20	92.02	98.84	111.74	122.48	131.07	139.98	149.03	156.77
14	87.10	94.40	100.78	114.21	124.44	133.22	142.28	151.47	159.33
13	89.92	97.11	104.32	118.18	128.93	138.60	148.02	157.58	165.76
12	93.90	101.41	108.52	123.56	135.37	143.97	153.75	163.69	172.13
11	.95.94	103.62	111.31	126.02	137.52	147.20	157.20	167.36	176.03
10	98.30	106.16	114.21	129.25	141.82	151.50	161.81	172.26	181.13
9	100.99	109.07	117.11	132.58	145.05	154.71	165.23	175.91	185.03
8	103.89	112.20	120.65	136.44	149.33	160.08	170.96	182.00	191.44
7	105.93	114.40	123.01	139.23	152.56	163.31	174.38	185.68	195.30
6	109.15	117.88	126.78	142.88	156.83	167.51	178.88	190.46	200.36
5	111.08	119.97	128.93	145.79	160.08	170.82	182.48	194.22	204.30
4	114.75	123.93	133.22	150.41	164.36	176.18	188.16	200.25	210.71
3	119.25	128.79	137.52	156.26	170.78	182.64	195.08	207.66	218.44
2	128.48	138.76	148.61	168.64	184.79	197.66	211.05	224.74	235.58
1	136.35	147.26	158.25	179.33	194.63	209.48	223.65	238.16	250.54
A	145.34	155.48	165.62	189.28	206.18	219.70	236.60	250.12	263.63
B	150.36	160.85	171.33	195.81	213.29	227.25	244.76	258.74	272.73
C	155.45	166.30	177.14	202.44	220.52	234.98	253.06	267.51	281.97
D	160.63	171.84	183.04	209.19	227.87	242.81	261.48	276.44	291.97
E	165.89	177.47	189.05	216.05	235.34	250.77	270.06	285.49	300.93
F	175.43	187.61	199.78	227.75	248.70	265.02	285.40	301.70	318.02

TABLE A-2 *continued*

Orifice Size (Decimal or DMS)	3	3.5	4	5	6	7	8	9	10
G	180.81	193.43	206.04	235.49	256.50	273.33	294.35	311.18	327.99
H	187.81	200.91	214.01	244.59	266.42	283.89	305.74	323.20	340.67
I	196.38	210.08	223.77	255.75	278.58	296.85	319.68	337.95	356.20
J	203.66	217.87	232.08	265.24	288.92	307.87	331.55	350.49	396.44
K	209.59	224.22	238.84	272.95	297.33	316.82	341.19	360.69	380.18
L	223.23	240.49	257.75	290.71	316.68	337.44	363.40	384.17	404.92
M	231.00	247.12	263.23	300.83	327.69	349.18	376.03	397.52	419.01
N	242.09	258.98	275.06	315.27	343.42	365.94	394.09	416.61	439.13
O	265.05	283.54	302.03	345.18	376.00	400.66	431.47	456.13	480.79
P	276.92	296.24	315.56	360.64	392.84	418.60	450.80	476.56	502.32
Q	292.57	312.98	333.39	381.03	415.05	442.26	476.28	503.49	530.71
R	305.03	326.33	347.62	397.26	432.73	461.10	496.59	524.95	553.32
S	321.45	344.01	366.57	418.62	456.01	485.91	523.28	553.19	583.09
T	340.19	363.92	387.65	443.04	482.59	514.24	553.79	585.44	617.09
U	359.46	383.54	409.61	468.14	509.93	543.36	585.17	618.60	652.04
V	377.26	403.58	429.90	491.31	535.17	570.27	614.14	649.23	684.33
W	395.48	423.07	450.66	515.05	561.04	597.83	643.82	680.60	717.39
X	418.34	447.53	476.72	544.82	593.47	632.39	681.03	719.94	758.86
Y	433.23	463.45	493.67	564.20	614.58	654.87	705.25	745.56	785.86
Z	452.75	484.33	515.91	589.64	642.26	684.38	737.02	779.14	821.26

Source: American Gas Association

TABLE A-3
Outside Design Temperatures for the United States

State & City	Winter DB	Summer DB	Daily Range	Summer WB	Latitude Deg.
ALABAMA					
Anniston	12	94	M	78	35
Birmingham	14	94	M	78	35
Gadsden	11	94	M	78	35
Mobile	21	93	L	80	30
Montgomery	18	95	M	78	30
Tuscaloosa	14	96	M	78	35
ALASKA					
Anchorage	-29	70	M	61	60
Barrow	-49	54	—	51	70
Bethel	-43	—	—	—	60
Cordova	-13	—	—	—	60
Fairbanks	-59	78	M	63	60
Juneau	-11	71	—	64	60
Ketchikan	4	—	—	—	55
Kodiak	4	66	—	60	55
Kotzebue	-46	—	—	—	65
Nome	-37	62	—	56	60
Seward	-4	—	—	—	60
Sitka	2	—	—	—	60
ARIZONA					
Bisbee	30	100	H	72	35
Flagstaff	-10	82	H	60	35
Globe	30	105	H	76	35
Nogales	15	98	H	71	30
Phoenix	25	106	H	76	35
Tucson	23	102	H	73	30
Winslow	2	95	H	65	35
Yuma	32	109	H	78	35
ARKANSAS					
Bentonville	0	95	M	76	35
Fort Smith	9	99	M	78	35
Hot Springs	12	97	M	78	35
Little Rock	13	96	M	79	35
Pine Bluff	14	96	M	80	35
Texarkana	16	97	M	79	35
CALIFORNIA					
Bakersfield	26	101	H	71	35
El Centro	26	109	H	80	35
Eureka	27	65	M	59	40
Fresno	25	99	H	72	35
Long Beach	31	84	M	70	35
Los Angeles	36	83	M	70	35
Montague	15	95	M	70	40
Needles	27	110	H	75	35
Oakland	30	81	M	63	40
Pasadena	31	93	M	70	35
Red Bluff	15	100	H	70	40
Sacramento	24	97	H	70	40
San Bernardino	26	98	H	73	35
San Diego	38	83	L	70	35
San Francisco	32	79	M	63	40
San Jose	30	88	M	67	35
COLORADO					
Boulder	-5	90	M	63	40
Colorado Springs	-9	88	H	62	40
Denver	-9	90	H	64	40
Durango	-10	86	H	63	35
Fort Collins	-18	89	M	62	40
Grand Junction	-2	94	H	63	40
Leadville	-18	73	M	55	40
Pueblo	-14	94	H	67	40
CONNECTICUT					
Bridgeport	-1	88	L	76	40
Hartford	-4	88	M	76	40
New Haven	0	86	M	76	40
New London	0	86	L	75	40
Norwalk	-5	89	L	76	40
Torrington	0	90	M	75	40
Waterbury	-5	88	M	76	40
DELAWARE					
Dover	8	90	M	78	40
Milford	10	90	M	78	40
Wilmington	6	90	M	77	40
DIST. OF COLUMBIA					
Washington	12	92	M	77	40
FLORIDA					
Apalachicola	25	95	L	80	30
Fort Myers	34	92	M	80	25
Gainesville	24	94	M	79	30
Jacksonville	26	94	M	79	30
Key West	50	89	L	79	25
Miami	39	90	L	79	25
Orlando	29	94	M	79	30
Pensacola	25	90	L	81	30
Tallahassee	21	94	M	79	30
Tampa	32	91	M	80	30
GEORGIA					
Athens	12	94	M	77	35
Atlanta	14	92	M	77	35
Augusta	17	95	M	79	35
Brunswick	24	95	L	80	30
Columbus	19	96	M	79	35
Macon	18	96	M	79	35
Rome	11	95	M	77	35
Savannah	21	94	M	80	30
Waycross	20	95	M	79	30
IDAHO					
Boise	0	93	H	66	45
Idaho Falls	-17	88	H	64	45
Lewiston	1	96	H	66	45
Pocatello	-12	91	H	63	45
Twin Falls	-5	94	H	64	40
ILLINOIS					
Aurora	-13	91	M	77	40
Bloomington	-7	92	M	78	40
Cairo	0	100	M	78	35
Champaign	-6	94	M	78	40
Chicago	-9	90	M	75	40
Danville	-6	94	M	78	40
Decatur	-6	93	M	78	40
Elgin	-14	90	M	76	40
Joliet	-11	92	M	77	40
Moline	-12	91	M	77	40
Peoria	-8	92	M	77	40
Rockford	-13	90	M	76	40
Rock Island	-10	95	M	76	40
Springfield	-7	92	M	78	40
Urbana	-10	95	M	77	40
INDIANA					
Elkhart	-10	95	M	75	40
Evansville	1	94	M	78	40
Fort Wayne	-5	91	M	76	40
Indianapolis	-5	91	M	77	40
Lafayette	-7	92	M	77	40
South Bend	-6	89	M	76	40
Terre Haute	-3	93	M	78	40
IOWA					
Burlington	-10	92	M	78	40
Cedar Rapids	-14	90	M	76	40
Charles City	-20	95	M	75	45
Clinton	-13	90	M	77	40
Council Bluffs	-14	94	M	78	40
Davenport	-10	95	M	78	40
Des Moines	-13	92	M	77	40
Dubuque	-17	90	M	76	40
Fort Dodge	-18	92	M	77	40
Keokuk	-9	93	M	78	40
Marshalltown	-16	91	M	77	40
Sioux City	-17	93	M	77	40
Waterloo	-18	89	M	76	40
KANSAS					
Atchison	-9	95	M	78	40
Concordia	-10	95	M	78	40
Dodge City	-5	97	H	73	40
Iola	-5	100	M	75	40
Leavenworth	-10	100	M	76	40
Salina	-4	99	M	76	40
Topeka	-4	96	M	78	40
Wichita	-1	99	M	76	40
KENTUCKY					
Bowling Green	1	95	M	78	35
Frankfort	0	95	M	78	40
Hopkinsville	4	95	M	78	35
Lexington	0	92	M	77	40
Louisville	1	93	M	78	40
Owensboro	0	94	M	78	40
Shelbyville	0	95	M	78	40
LOUISIANA					
Alexandria	20	95	M	80	30
Baton Rouge	22	94	M	80	30
New Orleans	29	91	M	80	30
Shreveport	18	96	M	80	30
MAINE					
Augusta	-13	86	L	73	45
Bangor	-14	85	L	73	45
Bar Harbor	-10	85	L	73	45
Belfast	-10	85	L	73	45
Eastport	-10	81	L	70	45
Lewiston	-14	86	L	73	45
Millinocket	-22	85	M	72	45
Orono	-20	85	M	70	45
Portland	-14	85	M	73	45
Presque Isle	-20	85	L	73	45
Rumford	-15	85	L	73	45
MARYLAND					
Annapolis	10	90	M	78	40
Baltimore	8	91	M	78	40
Cambridge	10	90	L	78	40
Cumberland	0	92	M	75	40
Frederick	2	92	M	77	40
Frostburg	-5	90	M	75	40
Salisbury	10	90	M	78	40
MASSACHUSETTS					
Amherst	-5	90	M	78	40
Boston	-1	88	M	74	40
Fall River	-1	86	L	74	40
Fitchburg	-5	90	M	75	40
Framingham	-7	89	M	75	40
Lawrence	-9	88	M	74	45
Lowell	-7	89	M	74	45
Nantucket	0	85	L	74	40
New Bedford	3	84	L	73	40
Pittsfield	-11	84	M	72	40
Plymouth	0	85	L	75	40
Springfield	-8	88	M	74	40
Worcester	-8	87	M	73	40
MICHIGAN					
Alpena	-11	85	M	73	45
Ann Arbor	-5	90	M	75	40
Big Rapids	-5	90	M	75	45
Cadillac	-10	90	M	75	45
Calumet	-20	80	M	73	45
Detroit	0	88	M	75	40
Escanaba	-13	80	M	71	45
Flint	-7	87	M	75	45
Grand Haven	-5	90	M	75	45
Grand Rapids	-3	89	M	74	45
Houghton	-20	80	M	73	45
Kalamazoo	-5	89	M	75	45
Lansing	-4	87	M	75	45
Ludington	-5	90	M	75	45
Marquette	-14	86	M	71	45
Muskegon	-2	85	M	74	45
Port Huron	-6	88	M	74	45
Saginaw	-7	86	M	75	45
Sault Ste. Marie	-18	81	M	71	45
MINNESOTA					
Alexandria	-26	88	M	74	45
Duluth	-25	82	M	71	45
Minneapolis	-19	89	M	75	45
Moorhead	-30	95	M	75	45
St. Cloud	-26	88	M	75	45
St. Paul	-19	89	M	75	45
MISSISSIPPI					
Biloxi	26	92	L	81	30
Corinth	13	95	M	79	35
Hattiesburg	18	95	M	79	30
Jackson	17	96	M	78	30
Meridian	15	95	M	79	30
Natchez	18	94	L	80	30
Vicksburg	13	96	L	79	30
MISSOURI					
Columbia	-4	95	M	78	40
Hannibal	-7	94	M	78	40
Kansas City	-2	97	M	77	40
Kirksville	-13	94	M	78	40
St. Joseph	-8	95	M	78	40
St. Louis	-2	95	M	78	40
Springfield	0	100	M	77	40
MONTANA					
Anaconda	-30	85	H	59	45
Billings	-19	91	H	66	45
Butte	-34	83	H	59	45
Great Falls	-29	88	H	63	50
Havre	-32	87	M	64	50
Helena	-27	87	H	63	45
Kalispell	-17	84	H	63	50
Miles City	-27	94	H	69	45
Missoula	-16	89	H	63	45
NEBRASKA					
Grand Island	-14	95	H	75	40
Hastings	-11	96	M	75	40
Lincoln	-10	96	M	77	40
Norfolk	-18	95	M	76	40
North Platte	-13	94	H	73	40
Omaha	-12	94	M	78	40
Valentine	-20	95	M	78	45
York	-15	95	M	78	40
NEVADA					
Elko	-21	92	H	62	40
Las Vegas	18	106	H	71	35
Reno	-2	92	H	62	40
Tonopah	2	92	M	63	40
Winnemucca	-8	95	H	62	40
NEW HAMPSHIRE					
Berlin	-25	85	H	71	45
Claremont	-19	87	H	73	45
Concord	-17	88	H	73	45
Franklin	-15	85	M	73	45
Hanover	-15	85	M	73	45
Keene	-17	88	H	73	45
Manchester	-11	89	M	74	45
Nashua	-10	85	L	74	45
Portsmouth	-8	86	L	73	45
NEW JERSEY					
Asbury Park	5	90	L	78	40
Atlantic City	10	88	L	77	40
Bayonne	0	90	L	76	40
Belvidere	0	90	M	75	40
Bloomfield	0	90	L	75	40
Bridgeton	5	90	L	78	40
Camden	5	90	L	78	40
East Orange	0	90	L	75	40
Elizabeth	0	90	L	75	40
Jersey City	0	90	L	75	40
Newark	6	91	M	76	40
New Brunswick	3	89	L	76	40
Paterson	3	91	L	76	40
Phillipsburg	1	91	M	76	40
Trenton	7	90	L	77	40

TABLE A-3 continued

State & City	Winter DB	Summer DB	Daily Range	Summer WB	Latitude Deg.
NEW MEXICO					
Albuquerque	6	94	M	65	35
El Morro	0	85	H	65	35
Raton	-11	90	H	65	35
Roswell	5	99	H	70	35
Santa Fe	-2	88	M	63	35
Tucumcari	1	97	H	70	35
NEW YORK					
Albany	-14	88	M	74	45
Auburn	-10	87	M	73	45
Binghamton	-8	89	M	72	40
Buffalo	-3	86	M	73	45
Canton	-20	85	M	73	45
Cortland	-11	88	M	73	45
Elmira	-5	90	M	73	40
Glens Falls	-17	86	M	72	45
Ithaca	-10	88	M	73	40
Jamestown	-5	86	M	73	40
Lake Placid	-15	90	M	73	45
New York	6	91	M	76	40
Niagara Falls	-2	86	M	74	45
Ogdensburg	-20	85	M	73	45
Oneonta	-13	87	M	72	45
Oswego	4	84	M	74	45
Port Jervis	0	90	L	75	40
Rochester	-5	88	M	74	45
Schenectady	-11	88	M	73	45
Syracuse	-10	87	M	74	45
Watertown	-20	84	M	74	45
NORTH CAROLINA					
Asheville	8	88	M	74	35
Charlotte	13	94	M	77	35
Greensboro	9	91	M	76	35
Hatteras	20	90	L	80	35
New Bern	14	92	L	80	35
Raleigh	13	92	M	78	35
Salisbury	10	90	M	78	35
Wilmington	19	91	M	81	35
Winston-Salem	9	91	M	76	35
NORTH DAKOTA					
Bismarck	-31	91	H	72	45
Devils Lake	-30	89	M	71	50
Dickinson	-31	93	H	70	45
Fargo	-28	88	H	74	45
Grand Forks	-30	87	M	72	50
Jamestown	-29	91	M	73	45
Minot	-31	88	M	70	50
Pembina	-35	90	M	73	50
Williston	-28	90	M	69	50
OHIO					
Akron	-5	87	M	73	40
Cincinnati	2	92	M	77	40
Cleveland	-2	89	M	75	40
Columbus	-1	88	M	76	40
Dayton	-2	90	M	75	40
Lima	-6	91	M	76	40
Marion	-5	91	M	76	40
Sandusky	-2	90	M	75	40
Toledo	-5	90	M	75	40
Warren	-6	88	M	74	40
Youngstown	-5	86	M	74	40
OKLAHOMA					
Ardmore	9	101	M	78	35
Bartlesville	-1	99	M	78	35
Guthrie	0	100	M	77	35
Muskogee	6	99	M	78	35
Oklahoma City	4	97	M	77	35
Tulsa	4	99	M	78	35
Waynoka	-5	105	M	75	35

State & City	Winter DB	Summer DB	Daily Range	Summer WB	Latitude Deg.
OREGON					
Arlington	5	95	M	68	45
Baker	-10	92	M	65	45
Eugene	16	88	H	67	45
Medford	15	94	H	68	40
Pendleton	-2	94	H	65	45
Port'and	17	85	M	67	45
Roseburg	19	91	H	67	45
Salem	15	88	H	67	45
Wamic	0	90	H	66	45
PENNSYLVANIA					
Altoona	-4	87	M	73	40
Bethlehem	0	90	M	75	40
Coatesville	5	90	M	75	40
Erie	1	85	M	74	40
Harrisburg	4	89	M	75	40
New Castle	-7	89	M	74	40
Oil City	-5	90	M	75	40
Philadelphia	7	90	M	77	40
Pittsburgh	-1	87	M	74	40
Reading	1	90	M	76	40
Scranton	-3	87	M	74	40
Warren	-8	87	M	73	40
Williamsport	4	90	M	76	40
York	-5	89	M	75	40
RHODE ISLAND					
Block Island	5	85	L	75	40
Bristol	0	90	L	75	40
Kingston	0	85	L	75	40
Pawtucket	0	90	M	75	40
Providence	0	86	M	75	40
SOUTH CAROLINA					
Charleston	19	92	L	80	35
Columbia	16	96	M	79	35
Florence	16	94	M	79	35
Greenville	14	93	M	76	35
Spartanburg	13	93	M	76	35
SOUTH DAKOTA					
Aberdeen	-29	92	M	75	45
Huron	-24	93	H	75	45
Pierre	-21	96	M	74	45
Rapid City	-17	94	H	71	45
Sioux Falls	-21	92	H	75	45
Watertown	-27	90	M	74	45
TENNESSEE					
Chattanooga	11	94	M	78	35
Jackson	8	95	M	79	35
Johnson City	0	95	M	78	35
Knoxville	9	92	M	76	35
Memphis	11	96	M	79	35
Nashville	6	95	M	78	35
TEXAS					
Abilene	12	99	M	75	30
Amarillo	2	96	H	71	35
Austin	19	98	M	78	30
Brownsville	32	92	M	80	25
Corpus Christi	28	93	M	80	30
Dallas	14	99	M	78	35
Del Rio	24	99	H	77	30
El Paso	16	98	M	69	30
Fort Worth	14	100	H	78	35
Galveston	28	89	L	81	30
Houston	23	94	M	80	30
Palestine	16	97	M	79	30
Port Arthur	25	92	M	80	30
San Antonio	22	97	M	77	30
Waco	16	99	M	78	30

State & City	Winter DB	Summer DB	Daily Range	Summer WB	Latitude Deg.
UTAH					
Logan	-7	91	H	65	40
Milford	-5	95	H	66	40
Ogden	-3	92	H	65	40
Salt Lake City	-2	94	H	66	40
VERMONT					
Bennington	-10	90	M	73	45
Burlington	-18	85	M	73	45
Montpelier	-20	90	M	73	45
Newport	-20	85	M	73	45
Northfield	-20	90	M	73	45
Rutland	-18	85	M	73	45
VIRGINIA					
Cape Henry	15	90	L	78	35
Charlottesville	7	90	M	77	40
Danville	9	92	M	78	35
Lynchburg	10	92	M	76	35
Norfolk	18	91	L	78	35
Petersburg	10	94	M	78	35
Richmond	10	93	M	78	40
Roanoke	9	91	M	76	35
Wytheville	5	90	M	76	35
WASHINGTON					
Aberdeen	19	80	L	60	45
Bellingham	8	82	L	65	50
Everett	13	78	L	65	50
North Head	20	80	L	65	50
Olympia	15	83	H	65	45
Seattle	17	80	M	65	50
Spokane	-5	90	H	64	50
Tacoma	14	81	M	66	45
Tatoosh Island	20	80	L	65	50
Walla Walla	5	96	H	68	45
Wenatchee	-2	92	M	66	50
Yakima	-1	92	H	67	45
WEST VIRGINIA					
Bluefield	1	86	M	73	35
Charleston	1	90	M	75	40
Elkins	-4	84	M	73	40
Fairmont	0	90	M	75	40
Huntington	4	93	M	76	40
Martinsburg	1	94	M	77	40
Parkersburg	2	91	M	76	40
Wheeling	0	89	M	75	40
WISCONSIN					
Ashland	-27	83	M	71	45
Beloit	-13	90	M	76	45
Eau Claire	-21	88	M	74	45
Green Bay	-16	85	M	73	45
La Crosse	-18	88	M	76	45
Madison	-13	88	M	75	45
Milwaukee	-11	87	M	75	45
Oshkosh	-20	90	M	75	45
Sheboygan	-20	90	M	75	45
WYOMING					
Casper	-20	90	H	62	45
Cheyenne	-15	86	H	62	40
Lander	-26	90	H	63	45
Sheridan	-21	92	H	65	45
Yellowstone Park	-35	85	H	62	45

The U.S. summer design temperatures represent the maximum wet and dry bulb temperatures recorded for each area reduced by 2½%. It is unusual for both wet and dry bulb temperatures to exceed the figures shown here simultaneously. The 2½% reduction is applied to the U.S. data for the months of June through September only.

The U.S. winter design temperatures shown represent the average of the coldest temperatures recorded for that area each year for approximately a twenty-five year period, often referred to as the Median of Extremes.

Source: ARI, Air Conditioning and Refrigeration Institute

TABLE A-4

Heat Transfer Factors Table

Item	30	35	40	45	50	55	60	65	70	75	80	85	90	95	100
WINDOWS, WOOD OR METAL FRAME (Btuh per sq ft) Factors include heat loss for transmission and infiltration @ 15 mph															
No. 1 Double-Hung, Horizontal-Slide, Casement or Awning (Infiltration less than 0.50 cfm/ft @ 25 mph certified by test)															
(a) Single glass	40	50	55	60	70	75	80	90	95	105	110	115	125	130	135
(b) With double glass or insulating glass	25	30	35	40	45	45	50	55	60	65	70	75	75	80	85
(c) With storm sash	25	25	30	35	40	45	45	50	55	60	60	65	70	75	80
No. 2 Double-Hung, Horizontal-Slide, Casement or Awning (Infiltration less than 0.75 cfm/ft @ 25 mph certified by test)															
(a) Single glass	45	50	60	65	75	80	90	95	105	110	120	125	135	140	150
(b) With double glass or insulating glass	30	35	40	45	50	55	60	65	70	75	80	80	85	90	95
(c) With storm sash	25	30	35	40	45	50	55	60	60	65	70	75	80	85	90
No. 3 Other Double-Hung, Horizontal-Slide, Casement or Awning															
(a) Single glass	75	90	105	115	130	140	155	165	180	195	205	220	230	245	255
(b) With double glass or insulating glass	60	70	80	90	105	115	125	135	145	155	165	175	185	195	205
(c) With storm sash	35	40	45	55	60	65	70	75	85	90	95	100	105	110	120
No. 4 Fixed or Picture															
(a) Single glass	40	50	55	60	70	75	85	90	95	105	110	115	125	130	140
(b) Double glass or with storm sash	25	30	35	40	45	45	50	55	60	65	70	75	75	80	85
No. 5 Jalousie (Infiltration less than 1.5 cfm/sq ft @ 25 mph certified by test)															
(a) Single glass	60	70	80	90	100	105	115	125	135	145	155	165	175	185	195
(b) With storm sash	35	40	50	55	60	65	70	75	80	85	90	100	105	110	115
No. 6 Other Jalousie															
(a) Single glass	225	265	300	340	375	415	450	490	525	565	600	640	675	715	750
(b) With storm sash	65	75	90	100	110	120	135	145	155	165	175	190	200	210	220
DOORS (Btuh per sq ft) Factors include heat loss for transmission and infiltration @ 15 mph															
No. 7 Sliding Glass Doors (Infiltration less than 1.0 cfm/sq ft @ 25 mph certified by test)															
(a) Single glass	50	60	70	75	85	95	100	110	115	125	135	140	150	160	165
(b) Double glass	40	45	50	55	60	65	70	80	90	95	100	105	115	120	125
No. 8 Other Sliding Glass Doors															
(a) Single glass	75	85	100	115	125	140	150	165	175	190	200	210	225	240	250
(b) Double glass	60	70	80	90	100	110	120	130	140	150	160	170	180	190	200
No. 9 Other Doors															
(a) Weatherstripped and with storm door	40	45	55	60	65	75	80	85	90	100	105	110	120	125	130
(b) Weatherstripped or with storm door	70	85	95	110	120	135	145	155	170	180	195	205	215	230	240
(c) No weatherstripping or storm door	135	160	180	200	225	250	270	290	315	340	360	380	405	430	450
WALLS AND PARTITIONS (Btuh per sq ft) **No. 10 Wood Frame with Sheathing and Siding, Veneer or Other Exterior Finish**															
(a) No insulation	8	9	10	11	13	14	15	16	18	19	20	21	23	24	25
(b) Expanded polystyrene extruded board sheathing (R-5)	3	4	4	5	6	6	7	7	8	8	9	9	10	10	11
(c) R-7 batt insulation (2"-2¾")	3	4	4	5	6	6	7	7	8	8	9	9	10	10	10
(d) R-11 batt insulation (3"-3½")	2	2	3	3	4	4	4	5	5	5	6	6	6	7	7
(e) R-13 batt insulation (3½"-3⅝")	2	2	2	3	3	3	4	4	4	5	5	5	5	6	6
No. 11 Partition between conditioned and unconditioned spaces															
(a) Finished one side only, no insulation	17	19	22	25	28	30	33	36	39	41	44	47	49	52	55
(b) Finished both sides, no insulation	9	11	12	14	16	17	19	20	22	23	25	26	28	29	31

Table header note: Design Temperature Difference, Degrees

* Insulation is produced in different densities and materials such as mineral fibers, glass fibers, etc.; therefore, there is some variation in thickness for the same R-value among various manufacturers' products.

TABLE A–4 *continued*

Item / No. 11 Cont'd	30	35	40	45	50	55	60	65	70	75	80	85	90	95	100
(c) Partition with 1" expanded polystyrene extruded board, R-5, finished both sides	4	4	5	5	6	7	7	8	8	9	10	10	11	11	12
(d) R-7 insulation (2"-2¾") finished both sides	3	4	4	5	5	6	6	7	7	8	8	9	9	10	10
(e) R-11 insulation (3"-3½") finished both sides	2	3	3	4	4	4	5	5	6	6	6	7	7	8	8
(f) R-13 insulation (3½"-3⅝") finished both sides	2	2	3	3	4	4	4	5	5	5	6	6	6	7	7
No. 12 Solid Masonry, Block or Brick															
(a) Plastered or plain	14	16	18	20	22	25	27	29	32	34	36	38	40	43	45
(b) Furred, no insulation	9	10	12	13	14	16	17	19	20	22	23	25	26	28	29
(c) Furred, with R-5 insulation (Nominal 1½")	4	5	5	6	6	7	8	8	9	10	10	11	12	12	13
No. 13 Basement or Crawl Space															
(a) Above grade, no insulation	15	18	20	23	26	28	31	33	36	38	41	43	46	48	51
(b) Wall of conditioned crawl space, R-3.57 insulation (molded bead bd.)	5	6	7	8	9	10	11	12	13	14	14	15	16	17	18
(c) Wall of conditioned crawl space, R-5 insulation (ext. polystyrene bd.)	4	5	6	6	7	8	9	9	10	11	12	12	13	14	14
(d) Wall of crawl space used as supply plenum, R-3.57 insulation (molded bead bd.)	11	12	13	14	15	16	16	17	18	19	20	21	22	23	24
(e) Wall of crawl space used as supply plenum, R-5 insulation (ext. polystyrene bd.)	9	10	10	11	12	12	13	14	15	15	16	17	17	18	19
(f) Below grade wall	2	2	2	3	3	3	4	4	4	5	5	5	5	6	6
CEILINGS AND ROOFS (Btuh per sq ft)															
No. 14 Ceiling Under Unconditioned Space or Vented Roof															
(a) No insulation	18	21	24	27	30	33	36	39	42	45	48	51	54	57	60
(b) R-11 insulation (3"-3¼")	2	3	3	4	4	4	5	5	6	6	6	7	7	8	8
(c) R-19 insulation (5¼"-6½")	2	2	2	2	2	3	3	3	4	4	4	4	4	5	5
(d) R-22 insulation (6"-7")	1	1	2	2	2	2	2	3	3	3	3	3	4	4	4
(e) Ceiling under unconditioned room	9	11	12	14	15	17	18	20	21	23	24	26	27	29	30
No. 15 Roof on Exposed Beams or Rafters															
(a) Roofing on 1½ in. wood decking, no insulation	10	12	14	15	17	19	20	22	24	26	27	29	31	32	34
(b) Roofing on 1½ in. wood decking, 1 in. insulation between roofing and decking	5	6	7	8	8	9	10	11	12	13	13	14	15	16	17
(c) Roofing on 1½ in. wood decking, 1½ in. insulation between roofing and decking	4	5	5	6	7	7	8	9	10	10	11	12	12	13	14
(d) Roofing on 2 in. coarse, shredded wood plank	6	7	8	9	10	11	12	14	15	16	17	18	19	20	21
(e) Roofing on 3 in. coarse, shredded wood plank	5	5	6	7	8	8	9	10	11	11	12	13	14	14	15
(f) Roofing on 1½ in. insulating fiberboard decking	6	7	8	9	10	10	11	12	13	14	15	16	17	18	19
(g) Roofing on 2 in. insulating fiberboard decking	4	5	6	7	8	8	9	10	11	11	12	13	14	14	15
(h) Roofing on 3 in. insulating fiberboard decking	3	4	4	5	6	6	7	7	8	8	9	9	10	10	11
No. 16 Roof-Ceiling Combination															
(a) No insulation	9	11	12	14	16	17	19	20	22	23	25	26	28	29	31
(b) R-11 insulation (3"-3½") between ceiling and decking	2	2	3	3	4	4	4	5	5	5	6	6	6	7	7
(c) R-19 insulation (5¼"-6½") between ceiling and decking	1	2	2	2	2	2	3	3	3	3	4	4	4	4	5
(d) R-22 insulation (6"-7")	1	1	2	2	2	2	2	3	3	3	3	3	4	4	4
FLOORS (Btuh per sq ft)															
No. 17 Floors Over Unconditioned Space															
(a) Over unconditioned room	4	5	6	6	7	8	8	9	10	11	11	12	13	13	14
(b) Over open or vented space or garage, no insulation	8	10	11	13	14	15	17	18	20	21	22	24	25	27	28
(c) Over open or vented space or garage, R-7 insulation (2"-2¾")	3	3	4	4	5	5	6	6	7	7	8	8	9	9	9
(d) Over open or vented space or garage, R-11 insulation (3"-3½")	2	2	3	3	4	4	4	5	5	5	6	6	6	7	7
(e) Over open or vented space or garage, R-19 insulation (5¼"-6½")	1	2	2	2	2	2	3	3	3	3	4	4	4	4	4
No. 18 Floor of Room Over Heated Basement or Crawl Space	0	0	0	0	0	0	0	0	0	0	0	0	0	0	0
No. 19 Basement Floor	1	1	1	1	1	2	2	2	2	2	3	3	3	3	3
FLOORS (Btuh per ft of Perimeter)															
No. 20 Concrete Slab Floor, Unheated															
(a) No edge insulation	35	40	40	45	45	50	50	55	55	60	60	65	65	70	75
(b) 1 in. edge insulation	25	30	30	35	35	40	40	45	45	50	50	55	55	60	60

Design Temperature Difference, Degrees

TABLE A–4 *continued*

Item No. 20 cont'd	Design Temperature Difference, Degrees														
	30	35	40	45	50	55	60	65	70	75	80	85	90	95	100
(c) 2 in. edge insulation	15	20	20	25	25	30	30	35	35	40	40	45	45	50	50
No. 21 Concrete Slab Floor with Perimeter System in Slab															
(a) No edge insulation	60	70	80	90	95	105	115	125	135	145	150	160	170	180	190
(b) 1 in. edge insulation	45	50	55	60	65	70	75	80	85	90	95	100	105	110	115
(c) 2 in. edge insulation	25	30	35	40	45	50	55	60	65	70	75	80	85	90	95
No. 22 Floor of Heated Crawl Space															
(a) Less than 18 in. below grade	35	40	40	45	45	50	50	55	55	60	60	65	65	70	75
(b) 18 in. or more below grade	15	20	20	25	25	30	30	35	35	40	40	45	45	50	50
No. 23 Floor of Crawl Space Used as Supply Plenum															
(a) Less than 18 in. below grade	45	50	50	60	60	65	65	70	70	80	80	85	85	90	95
(b) 18 in. or more below grade	20	25	25	30	30	40	40	45	45	50	50	60	60	65	65
OUTSIDE AIR (Btuh per cfm)															
No. 24 Ventilation or Makeup Air	30	40	45	50	55	60	65	70	75	80	85	90	95	105	110

Source: Reprinted from *Manual J* with permission of ACCA, Air Conditioning Contractors of America

TABLE A-5

Abbreviated Heat Transfer Factors Table

DESIGN TEMPERATURE DIFFERENCE, DEGREES	30	40	50	60	70	80	90	100	U
WINDOWS									
GLASS									
a) no storm sash	34	45	57	68	79	90	102	113	1.13
b) w/storm sash	23	30	38	45	53	60	68	75	0.75
c) double glass	18	24	31	37	43	49	55	61	0.61
DOORS									
a) no storm door	15	20	25	30	35	40	45	50	0.50
b) w/storm door	10	13	16	19	22	26	29	32	0.32
WALLS									
FRAME OR MASONRY									
a) no insulation	9	12	15	18	21	24	27	30	0.30
b) 2" insulation	4	6	7	8	10	11	13	14	0.14
c) more than 2" insulation	3	4	5	5	6	7	8	9	0.09
CEILINGS & ROOFS									
CEILINGS									
a) uninsulated	10	13	16	19	22	26	29	32	0.32
b) 2" insulation	4	5	6	7	8	10	11	12	0.12
c) 4" insulation	3	4	5	6	7	8	9	10	0.10
d) more than 4" insulation	2	3	4	5	6	6	7	8	0.08
FLOORS									
UNCONDITIONED ROOMS	4	6	7	8	10	11	13	14	0.14
CONCRETE SLAB, UNHEATED									
a) no insulation	5	7	9	10	12	14	15	17	0.17
CONCRETE, USING PERIMETER SYSTEM IN SLAB									
a) no insulation	57	76	95	114	133	152	181	190	1.90
INFILTRATION									
WINDOWS									
a) average fit, weatherstripped or w/storm sash	18	24	30	36	42	48	54	60	0.60
DOORS									
a) average fit	60	80	100	120	140	160	180	200	2.00
b) average fit, weatherstripped or w/storm door	30	40	50	60	70	80	90	100	1.00

Source: Lennox Industries Inc.

TABLE A–6
Short Form Calculation Sheet Used for Figuring Shell Load

JOB NAME _____ DATE _____

ADDRESS _____

DEALER _____

O.S. Design Temp _____ I.S. Design Temp _____
Design Temp Diff _____
Wall Construction _____ Insulation _____" Window Type _____
Roof Construction _____ Insulation _____" Floor _____

	AREA	HTF	BTUH LOSS	
GROSS WALL		✕	✕	
WINDOWS				
DOORS				
NET WALL				
CEILING				
FLOOR				
INFILTRATION W D	CRACKAGE			
TOTAL LOSS	✕	✕		

Source: Lennox Industries Inc.

TABLE A-7
Long Form Calculation Sheet Used for Figuring Room-by-Room Load

HEATING CONTRACTOR _____ ADDRESS _____

NAME OF JOB _____ ADDRESS _____

JOB NO. _____
DATE _____ BY _____
OUTSIDE DESIGN TEMP. _____
INSIDE DESIGN TEMP. _____
DESIGN TEMP. DIFF. _____

				1.		2.		3.		4.		5.		6.		7.		8.		9.	
	Table No.	"U" Factor	Multi-plier	Area Crack Cfm	Btuh Loss	Area Crack Cfm	Btuh Loss	Area Crack Cfm	Btuh Loss	Area Crack Cfm	Btuh Loss	Area Crack Cfm	Btuh Loss	Area Crack Cfm	Btuh Loss	Area Crack Cfm	Btuh Loss	Area Crack Cfm	Btuh Loss	Area Crack Cfm	Btuh Loss
1. Room Use and Floor Level																			
2. Room Length and Width																			
3. Running Ft of Exposed Wall																			
4. Ceiling Height																			
5. Exposure Type																			
6. Gross Exposed Wall																			
7. and Windows Doors																			
8. Net Exposed Wall																			
9. Cold Partition																			
10. Cold Ceiling																			
11. Cold Floor																			
12. Infiltration, (Leakage) or Air Change																			
13. Room Sub-Total Btuh																			
14. Btuh Adjustment																			
15. Room Total Btuh																			
16. Cfm @ Temp. Rise																			
17. Adjusted cfm																			

Source: Lennox Industries Inc.

TABLE A–8
Equal Friction Chart

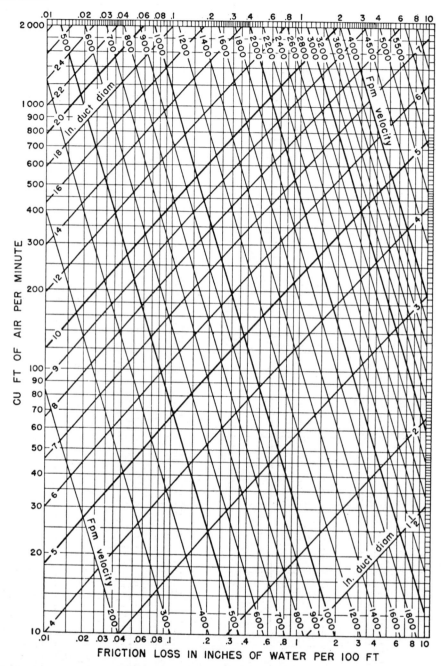

(Based on Standard Air of 0.075 lb per cu ft density flowing through average, clean, round, galvanized metal ducts having approximately 40 joints per 100 ft.) Caution: Do not extrapolate below chart.

Source: Reprinted with permission from American Society of Heating, Refrigerating and Air-Conditioning Engineers, Inc., *ASHRAE Handbook, 1977 Fundamentals* (New York, 1977)

TABLE A-9
Rectangular Duct Equivalent

Duct Diameter, in.	Equivalent Rectangular Stack and Duct Sizes, in.												
4.0		4.5×3											
4.5	8×2.25	6×3											
5.0	10×2.25		8×3.25	5×4	4×5			3×8					
5.5	12×2.25		10×3.25	7×4	5×5			4×8					
6.0	14×2.25		10×3.25	8×4	6×5			4×8					
6.5			12×3.25	9×4	7×5	6×6		5×8					
7.0			14×3.25	11×4	8×5	7×6		5×8					
7.5				13×4	10×5	8×6	7×7	6×8					
8.0				15×4	11×5	9×6	8×7	7×8					
8.5				17×4	13×5	10×6	9×7	8×8					
9.0				20×4	15×5	12×6	10×7	8×8					
9.5				22×4	17×5	13×6	11×7	9×8	8×9				
10.0				25×4	19×5	15×6	12×7	11×8	9×9				
10.5					21×5	16×6	14×7	12×8	10×9	9×10			
11.0					23×5	18×6	15×7	13×8	11×9	10×10			
11.5					26×5	20×6	17×7	14×8	12×9	11×10			
12.0					29×5	22×6	18×7	16×8	14×9	12×10			
12.5					32×5	24×6	20×7	17×8	15×9	13×10	11×12		
13.0					35×5	27×6	22×7	18×8	16×9	14×10	12×12		
13.5						30×6	24×7	20×8	17×9	16×10	13×12		
14.0						32×6	26×7	22×8	19×9	17×10	14×12		
14.5						35×6	28×7	24×8	20×9	18×10	15×12		
15.0						38×6	31×7	26×8	22×9	19×10	16×12	14×14	
15.5						41×6	33×7	28×8	24×9	21×10	17×12	14×14	
16.0						45×6	36×7	30×8	25×9	22×10	18×12	15×14	
16.5							38×7	32×8	27×9	24×10	19×12	16×14	
17.0							41×7	34×8	29×9	25×10	21×12	17×14	15×16
17.5							44×7	37×8	31×9	27×10	22×12	18×14	16×16
18.0								39×8	33×9	29×10	23×12	20×14	17×16
18.5								42×8	36×9	31×10	25×12	21×14	18×16
19.0								45×8	38×9	33×10	26×12	22×14	19×16
19.5								47×8	41×9	35×10	28×12	23×14	20×16
20.0								51×8	43×9	37×10	29×12	25×14	21×16

TABLE A-10

Various Ductwork Fittings with Equivalent Lengths

GROUP 1. SUPPLY AND RETURN. AIR TAKE-OFF PLENUM FITTINGS
(These fittings may also be installed on plenums for counter flow units.)

A = 10 ft   B = 10 ft   C = 10 ft   D = 15 ft   E = 35 ft   F = 35 ft   G = 45 ft   H = 45 ft

I = 65 ft   J = 45 ft   K = 65 ft   L = 65 ft

Not Recommended for Supply   Return Air Only

GROUP 2. REDUCING TRUNK DUCT FITTINGS

A = 15 ft   B = 30 ft   C = 20 ft   D = 60 ft   E = 15 ft   F = 75 ft   G = 35 ft   H = 35 ft   I = 15 ft   J = 60 ft   K = 20 ft   L = 30 ft

M = 25 ft   N = 40 ft   O = 25 ft   P = 50 ft   Q = 75 ft   R = 20 ft   S = 40 ft   T = 10 ft   U = 35 ft

NOTE: D – 3" Throat Radius
       F – No Radius
       Q – Not Recommended

TABLE A–10 *continued* _____

GROUP 3. EXTENDED PLENUM FITTINGS

*A ·40 ft

*C ·40 ft

*B ·15 ft

E· 5 ft

*F ·50 ft

*D ·15 ft

*G ·35 ft

Add 25 equiv. feet to each of
the 3 fittings nearest the unit
in each trunk duct and after
each reduction As shown by
asterisks.

GROUP 4   ROUND TRUNK DUCT FITTINGS
(Add 25 equivalent feet to each of the 3 fittings nearest the unit in each
Trunk Duct)

10 ft

25 ft

50 ft

35 ft

60 ft

30 ft

50 ft

35 ft

5 ft

50 ft

35 ft

TABLE A–10 *continued*

GROUP 5. ANGLES AND ELBOWS FOR TRUNK DUCTS
(Inside Radius = ½ Width of Duct)

A –

Trunk Width Inches	
4 to 15	5 ft
16 to 27	10 ft
28 to 41	15 ft
42 to 52	20 ft
53 to 64	25 ft

B –

Trunk Width Inches	
4 to 11	10 ft
12 to 21	15 ft
22 to 27	20 ft
28 to 33	25 ft
34 to 42	30 ft
43 to 51	40 ft
52 to 64	50 ft

C –

Trunk Width Inches	
4 to 6	20 ft
7 to 11	40 ft
12 to 15	55 ft
16 to 21	75 ft
22 to 27	100 ft
28 to 33	125 ft
34 to 42	150 ft

D –

Trunk Width Inches	
4 to 11	15 ft
12 to 21	20 ft
22 to 27	25 ft
28 to 42	40 ft

E = 5 ft   F = 10 ft   G = 30 ft   H = 15 ft   I = 30 ft

GROUP 6. ANGLES AND ELBOWS FOR INDIVIDUAL AND BRANCH DUCTS
(Inside Radius for A" and "B" = 3 in. and for "F" and "G" = 5 in.)

A = 5 ft   B = 10 ft   C = 25 ft   D = 5 ft   E = 10 ft

F = 5 ft

G –

10 in. wide	10 ft
12	15 ft
14	15 ft

H –

10 in. wide	40 ft
12	55 ft
14	55 ft

I –

3¼ in. X 10 in.	60 ft
12	75 ft
14	75 ft

J –

3¼ in. X 10 in.	75 ft
12	90 ft
14	90 ft

TABLE A–10 *continued*

K=125 ft   L=35 ft   M=10 ft   N=95 ft

O=25 ft   P=15 ft   Q=25 ft

Return Air Liner To Duct.   Stud Space To Liner.   Stud Space To Liner

GROUP 7. BOOT FITTINGS
(These values may also be used for floor Diffuser Boxes)

A=30 ft   B=35 ft   C=60 ft   D=55 ft   E=70 ft

F=45 ft   G=30 ft   H=50 ft   I=5 ft   J=15 ft

K=30 ft   L=30 ft   M=5 ft   N=15 ft   O=5 ft   P=5 ft

Source: Reprinted from National Environmental Systems Contractors Association, *Equipment Selection and System Design Procedures* (Arlington, VA, 1973) with permission from ACCA, Air Conditioning Contractors of America

INDEX